Great Plains Cattle Empire

Great Plains Cattle Empire
Thatcher Brothers and Associates (1875-1945)

Paul E. Patterson
Joy Poole

Texas Tech University Press

This book is typeset in Aldine 741 and Casa Blanca Antique. The paper used in this book meets the minimum requirements of ANSI/NISO Z39.48–1992 (R1997). ∞

Interior design by Melissa Bartz
Cover design by Lindsay Starr

Front cover photo courtesy of Montana Historical Society, Helena. Mahlon and John A. Thatcher photos courtesy of Pueblo City-County Library District.

Library of Congress Cataloging-in-Publication Data

Patterson, Paul E., 1926-
  Great Plains cattle empire : Thatcher Brothers and Associates (1875-1945) / Paul E. Patterson, Joy Poole.
        p.   cm.
  Includes bibliographical references (p.   ) and index.
  ISBN 0-89672-397-6 (cloth : alk. paper) ISBN 0-89672-563-4 (pbk : alk. paper)
  1. Cattle trade—Great Plains—History. 2. Great Plains—History. I. Poole, Joy. II. Title.
  HD9433.U5 G747 2000
  338.1'762'00978—dc21

                                                                                99-051009

ISBN-13 978-089672-397-9 (cloth)
ISBN-13 978-089672-563-8 (paper)

05 06 07 08 09 10 11 12 13 / 9 8 7 6 5 4 3 2 1
Printed in the United States of America
SHER

Texas Tech University Press
Box 41037
Lubbock, Texas 79409–1037 USA
800.832.4042
ttup@ttu.edu
www.ttup.ttu.edu

CONTENTS

Dedicated to my wife and best friend, Anndy, for being understanding when playing second fiddle to a word processor and for her continued encouragement.

P.E.P.

To the Diamond A and Circle Diamond cowboys.

J.L.P.

FOREWORD

An oldtime cowman told me once that the typical cattle market cycle has three stages. First, he said, is the "cowboy" stage, when times are tough, the market is at low ebb, and no one is in the business except hardy dyed-in-the-wool cattle people. Second comes the "trader" stage, when the markets begin to rise and potential profits can be visualized by astute speculators who know cattle and the livestock industry.

As prosperity continues, the cycle moves into its final "drugstore cowboy" stage, he said. This is when drugstore owners, doctors, lawyers, and other urban dwellers see ranchers and traders making money. They decide to get into the cattle business and become rich. They buy a pair of cowboy boots and wade in up to their chins, eventually saturating the market to the breaking point. After losing their shirts—and the boots—they liquidate, and the cycle drops back to the "cowboy" stage. Everything starts over again.

Though simplistic, the oldtimer's observation is basically valid as demonstrated in the Great Plains cattle-ranching history outlined by Paul Patterson and Joy Poole. Time and again, the prospect of big profits has lured investors into the industry, tantalized them a while with good returns, then cruelly smashed their hopes and dreams. Falling markets, droughts and paralyzing winters have wreaked havoc on cattle people over and over, yet always those good managers who survive find ways to get back on their feet for another round. These are the hands-on operators who know the land, the livestock and the markets, as opposed to the boardroom and windshield speculators who see the cow mainly as a figure on a ledger sheet.

Patterson and Poole show us both kinds. Their narrative is a veritable Who's Who of pioneer cattlemen. It stresses that the cattle industry is first of all a demanding business fraught with risk for the unwary. The ranges are littered with the bones of ill-conceived and loosely-managed speculative ventures.

Yet there have always been those whose skill and strong will allowed them to endure the periodic catastrophes that ruined so many of their contemporaries. Much of the book revolves around John Albert Thatcher and his brother Mahlon, who started their cattle empire on open range in Colorado and eventually saw it extend from Texas to the Dakotas, Montana, and Canada. Other notable industry pioneers who track across these pages include Henry W. Cresswell, Charles Goodnight, Robert O. Anderson, and Tony Day, to mention a few.

One point comes through clearly: there is a season for all things, but that season does not last forever. The one constant in life is change. Great herds of bison dominated the plains for thousands of years yet were virtually annihilated in a decade. The various Indian tribes had their time upon the land, then were pushed aside by the unstoppable tide of Euro-American settlement. The large 1870s–1880s cattle operations, most representing huge investments of Eastern and European capital, rose briefly, then fell to the relentless pressures of land-seeking homesteaders, money panics, overstocking, and disastrous winters.

The cattle industry today remains much as it was then, short periods of prosperity punctuated by longer stretches of break-even or financial loss. It continues to be challenged by recurring droughts and sagging markets. In public-lands states, the grazing of federally-owned acreage is under constant attack by environmentalists and other who seek to convert it to their own uses. The outlook for the American cattle industry is as clouded today as it has ever been in the past.

In giving us a detailed study of its history, Patterson and Poole offer us a perspective from which to view its present and perhaps acquire some inkling of its future.

Elmer Kelton

PREFACE

"Speakin' of cowpunchers," says Rawhide Rawlins, "I'm glad to see in the last few years that them that know the business have been writin' about 'em. It begin to look like they'd be wiped out without a history."

Charles Russell, *Trails Plowed Under*

The Great Plains is an elliptical, midcontinental basin that stretches from the northern prairie provinces in Canada south to the Chihuahuan Desert in Mexico. The area is bounded on the west by the evening shadows of the Rocky Mountains and on the east by the hundredth meridian.

The region's surface is characterized by deep, loamy soils, shifting sands, tilted strata, and heterogenous aggregates. This mantle has been laid down on igneous bedrock over the ages through the actions of eruptions, erosion, glaciers, and inland seas. Alpine ranges, rocky ridges, and solitary mesas were thrust above the rolling prairie eons ago by spasms and convolutions of the Earth's crust. The cutting edges of wind, the force of torrential rains, and the power of freezing and thawing over centuries has sculpted the contours of the Great Plains into a diverse landscape. Elevations range from thirty-five hundred to twelve thousand feet. Rocky pinnacles shadow sandy deserts and forested mountains rise above expanses of grasslands that turn verdant under the gentle coaxing of sun and rain or lie browned by drought or the killing cold of winter.

When the Earth was undergoing a warming trend and the massive sheet of ice that covered much of North America began to retreat, plant seeds that had lain dormant in the frozen soil for centuries began to stir under the rays of the sun. For centuries the ground cover of the Great Plains has supported a variety of animal life. Quadrupeds have evolved and flourished from the age of the dinosaurs—from the eohippus and saber-toothed tiger to the bison and its contemporaries. Most of the prehistoric beasts have faded into extinction. Only their tracks enduring in ancient sandstone beds and their fossilized bones embedded in strata of gray limestone provide proof of their existence.

Of all the large animals that have roamed over the Great Plains, none proved more adaptable and prolific than the bison. This herbivore, which has been known to reproduce at thirty years of age, multiplied almost without hindrance. That is until a meat-eating predator carrying fire and stone-tipped weapons came across the Bering Strait, the landmass that connected Asia and the American continent.

Lewis and Clark and other early explorers of the region west of the Missouri River talked about vast herds of bison. They told of a sea of shaggy brown that moved from morning until dusk like a rolling tide and stretched from horizon to horizon. Estimates, with about as much probability of being accurate as a count of the stars on a clear night, put the population of buffalo ranging on the Great Plains in 1867 at fifty million.[1] The bison, an ungainly, big-headed, pinch-rumped, humpbacked monarch of the prairie, was a renewing, self-sustaining horn of plenty for the first Americans who roamed the Great Plains. There was little or no waste from the carcass. Women tanned the hide for clothing and to cover the frame of the tepee. The fleshy parts, the viscera, and the marrow were used for food. Implements were fashioned from the bones, brains were used as a tanning agent, and hooves boiled down to make glue. Even the switch of the tail was used as a fly swatter and dung as fuel for fires.

After the man with hair on his pale face moved onto the land of the Plains Indians, buffalo hide became an important trade commodity. A tanned hide taken to a trader was the passkey to the white man's cornucopia of addictive wonders: sugar, flour, coffee, tobacco, iron pots and implements, trinkets, blankets, munitions, and liquor. By the mid-1840s, the Indians were killing in excess of two hundred thousand buffalo a year for their own use and for trade. By about 1870, tanneries were able to handle not only pelts of bison cows but the thicker, heavier ones of mature bulls. The popular use of buffalo hides for lap robes, floor coverings, and leather goods created a growing demand. A flint bullhide would bring as much as $2.[2] A good hunter, aided by a couple of skinners, could harvest as many as 150 head on an average day. Men set adrift in the aftermath of the Civil War swarmed to the slaughter.

The Sharps rifle, the grim reaper that threw a .50 caliber lead slug capable of downing a twelve-hundred-pound buffalo three-quarters of a mile away, sounded westward over the grassy hills and shallow swales of the frontier. By 1872, railroads were hauling nearly a million buffalo pelts a year to tanneries back East. The plains of Kansas and Nebraska were dotted with rotting carcasses, and the prevailing southwest winds carried clouds of blowflies and the stench of putrefying flesh.

The intensity of the killing began to have a telling effect. By 1879, commercial hide hunting had all but eliminated the southern herd of buffalo that ranged along and south of the Arkansas River in the land of the Comanche and Kiowas. The army of hunters was forced to turn to the northern herd and began reaping their grisly harvest near the Republican River in the land of the Cheyenne and Arapahoes. The age-old migratory habit of the herds was disrupted. The rather docile, slow-witted beasts became wary of the scent of the ubiquitous two-legged predator and the booming noise that brought the electrifying smell of blood to the herd. Wholesale slaughter by hunters continued, making it ever more difficult for the Indians with their stone-tipped weapons or small caliber rifles to make a kill.

Again it was a time of transition on the Great Plains. The Ice Age had come and gone. Now came the age when the bison, and the descendants of the people who had crossed over the Bering Strait, had proliferated and roamed as freely as the wind for some forty thousand years was coming to an end. Scavengers were gathering the remnant of the buffalo, piling huge ricks of bleached bones at railroad sidings. The indigenous first Americans faced annihilation from warfare and alien diseases brought by newcomers from Europe. Survivors were being subjugated and corraled onto reservations. Only the spirits of their ancestors still roamed unbridled where once "The People," as they called themselves, had lived in harmony with Mother Earth.

In 1865, the large herds of cattle and sheep were located mainly on ranches in California and Texas. Those animals were the proliferation from stock brought to the American continent on galleons from Spain in the fifteenth century. Production of livestock east of the Missouri River was limited to farm flocks, the seed stock coming with immigrants from Europe. Texans were learning the art of handling cattle on the open range. They copied the style of dress and horse tack from the *vaquero*, the mounted man who herded longhorns in Mexico.

By the terms of the Treaty of Guadalupe-Hidalgo, ending the Mexican War, the United States acquired all the land north of the Rio Grande that had been part of Mexico. The bulk of this vast acquisition was declared public domain. The next five decades saw the growth of entrepreneurs called cattlemen, hardy mounted men called cowboys, and speculating financiers. Future generations would learn about mounted men driving herds of cattle numbering in the thousands over hundreds of miles of lonesome, windswept prairie and across wide and treacherous rivers. The saga would tell of rip-roaring towns that sprouted and boomed at the intersections of railroad tracks which came from the east and meandering, rutted cattle trails from Texas. Frontier cowtowns flourished, like Abilene and Dodge City, where board-and-batten buildings flanked a dusty or muddy avenue. The typical main thoroughfare stretched from the cemetery on the hill to the railroad shipping pens on the prairie. And when trail-weary young cowpokes with money burning a hole in their pockets loosened their cinches and relieved pent-up cravings during a few days and nights of riotous living, it was like Sodom and Gomorrah revisited.

The history of the Great Plains, covering the relatively short span of years when it was predominately public domain, describes a way of life that had never been before: it came into being as a child of transition, proliferated like prairie grasses in a wet spring, and faded into yellowed tintypes and dusty memories. Winter storms, searing drought, overgrazing, the extreme fluctuations of the market, and the sodbuster with his plow and barbed wire were playing the death march to the days of the so-called cattle baron by the end of the nineteenth century.

The day of the speculating entrepreneur, the trail driver, the cowboy, and the great herds of cattle roaming over public domain was a never-before and a never-again time in the annals of ranching in the United States. Of the untold millions of cattle, sheep, and horses that roamed over the 140 million acres of prairie grasses blanketing the Great Plains in the 1890s, only a small fraction graze on the two percent of native sod that remains today.[3]

The brothers John A. and Mahlon D. Thatcher, from their base in the First National Bank of Pueblo, Colorado, were pioneers in merchandising, mining, and banking. Together with their associates Frank G. Bloom, Henry W. Cresswell, O. H. Perry Baxter, William E. Anderson, Burton C. Mossman, A. J. Day, and Mahlon T. Everhart, they led the development of the ranching industry on the Great Plains from 1871 to 1944.

At the peak of their cattle empire, the Thatcher brothers financed and directed the Bloom Land and Cattle Company, Cresswell Land and Cattle Company, Diamond A Cattle Company, and the Hatchet Cattle Company. Their herds of cattle, horses, and sheep, branded with the Circle Diamond, Diamond A, Bar CC, Tee Down Bar, Turkey Track, Hatchet, or Stroke Box, ranged on some eleven million acres of public and deeded lands from the southwest corner of New Mexico to the Whitemud River in the Provisional District of Assiniboia, Canada.

In the last half of the nineteenth century, a cowman's range was not reckoned in acres, but in claims called "range rights." Range rights were determined by the precept that "possession is nine-tenths of the law," and on the basis of "first come, first served." Rights and enforcement to the unwritten law went to the powerful. A cattleman could claim range rights on public domain over all the area he reasonably expected his branded cattle to utilize, and his hired cowboys, prowling the range horseback, could warrant. Stray cattle were usually driven off and encroaching sheep were often killed. Established range rights could be bartered and sold, not in a legal sense as could patented land, but just as acceptable by men of the open range.

ACKNOWLEDGMENTS

I would like to thank Mahlon T. White and Bob Johnston of Pueblo, Colorado, for their wager that the story would be told; John Korber of Pueblo for suffering through the early drafts and offering suggestions; and Alvin Davis of Lubbock, Texas, for his positive critique.

Paul E. Patterson

My love for this project started early in my career when I was Administrator of the Trinidad History Museum. A tour guide at the museum used to say, "Frank Bloom could start from his ranch on the border of Old Mexico and ride all the way to Canada, and never leave his property." At the time it seemed an unbelievable exaggeration. Now, after twenty years of researching Bloom's cattle ranching empire, I know it was damn near true.

During my research for this book, I had the real pleasure of meeting some colorful old Bloom cowboys like "Uncle" Joe Lopez, who became a dear friend and regular visitor, often arriving with his white Stetson in one hand and a box of chocolates in the other. Darwin Daniels and Theo Berlier, also famous Bloom cowboys, had wonderful stories of their annual roundups, cattle shipments, and days on the range. From them I learned that the Bloom and Thatcher companies were famous for "lots of horses and good food" that always attracted the best cowboys.

I am indebted to my friend and colleague Byron Price for his gentle nudging on this project. Byron told me during the first Santa Fe Trail Symposium in Trinidad of an original interview of Frank Bloom to be found at the J. E. Haley Memorial Library, which ultimately led me to the Bloom Land and Cattle Board Minutes. Bob Johnston Jr. located the minutes after he personally sorted through stacks of ledgers buried deep in the basement bank vaults of the First National Bank. Bob and his wife Doris opened their home and their wallets, providing support at critical junctures in the research and lending credence to the importance of the Bloom and Thatcher ranching empires. Many hours of laughter were shared with Richard and Willard Louden of Branson, Colorado, who did their best to educate this Iowa farm girl on the nuances of western cattle ranching.

Thanks to the following persons and organizations who assisted with the preparation of this study: Alberta Iliff Shattuck, who shared the stories of her mother, Alberta Bloom Iliff; Austin Hoover of the Rio Grande Historical Collections, Las Cruces, New Mexico; Phelps White of Roswell, New Mexico; the late Mahlon Everhart III of Hachita, New Mexico; Glen Aultman, photographer, of Trinidad, Colorado; the Arthur Roy Mitchell Museum, Trinidad, Colorado; Sallie Monroe Hall of Trinidad, Colorado; Ed Broadhead and other members of the Pueblo County Historical Society, Pueblo, Colorado; The Thatcher Families, Pueblo, Colorado; Jim DeMersman, former Director of the Rosemount Mansion and Thatcher archives, Pueblo, Colorado; Howard Munsell of Fowler, Colorado; the J. E. Haley Memorial Library, Midland, Texas; Dorothea Simonson, retired assistant Reference Librarian of the Montana Historical Society, Helena, Montana; the Phillips County Historical Society, Malta, Montana; The Glenbow Museum, Calgary, Canada; the Panhandle-Plains Museum, Canyon, Texas; the South Dakota State Historical Society, Pierre, South Dakota; and the entire library and manuscript staff of the Colorado Historical Society. Thanks also to Carole Young, formerly of Texas Tech University Press, for her perseverance.

Joy L. Poole

GO WEST YOUNG MAN

If any man is about to commence in the world, with little in his circumstance to prepossess him in favor of one section above another, we say to him publicly and privately, "Go to the West"; there your capacities are sure to be appreciated and your industry and energy rewarded.

Horace Greeley

John Albert Thatcher was born in the Pfoutz Valley of Pennsylvania on August 25, 1836. His parents were Henry and Lydia Thatcher. As a young man, John worked in his father's mercantile store in Martinsburg. He completed secondary schooling but declined his father's offer to finance the furthering of his education. John left the store and turned to teaching. He taught primary French in a one-room school at Lorberry in Schuylkill County, Pennsylvania.[1]

Articles in eastern publications about the world west of the Mississippi intrigued John. The articles told of adventures in a raw, wide-open country; of wild Indians and great herds of buffalo, elk, and pronghorn roaming over the limitless prairie; of bonanzas of gold and silver; and of fortunes being made in trade goods. These tales fueled the young man's imagination and fired his ambitions.

The pursuit of teaching children in the staid farming valleys of Pennsylvania paled in comparison to the challenge and the opportunities that beckoned from the virgin frontier. In 1857, the call of that vast land beyond the western horizon became irresistible. John walked to the blackboard and wrote: *"Pour prendre congé"* (I am going to take my leave).[2] He packed his belongings in a steamer trunk, bade his family farewell, and boarded a train headed west. He clerked in Missouri for five years until he could no longer resist the urge to join the flood of emigrants heading west.[3] The Kansas Pacific Railroad carried him to the end of the line at Fort Leavenworth, Kansas. He got a job as a teamster driving a team of oxen pulling a covered wagon in a caravan. The wagon creaked into Denver, Colorado Territory, in September 1862.

John's first job was in a tannery in Central City a few miles west of Denver. After several months at that malodorous work he returned to Denver where he took a job as a clerk in a general mercantile owned and operated by James Voorhees. Because people were traveling over a hundred miles from Pueblo to Denver and back to buy supplies, Voorhees decided to test the market in Pueblo. He sent John with a wagon load of goods to Pueblo. John would receive half of the profits in return for his labors.[4]

The early settlers called Pueblo Fountain City. The village comprised some mud and log shelters occupied by a few Anglo prospectors and a handful of Mexican farmers and sheep ranchers who clustered together for protection against the Apaches and Utes. Prospecting and mining for gold and silver on the eastern slopes of the Rocky Mountains was at a fever pitch, and the whole region along the foothills of the Rockies was as active as a busted anthill. By the summer of 1860, Fountain City had been renamed Pueblo, the Mexicans' word for village, and the population was growing.

It took eight days for a team of mules to pull John and his wagon load of merchandise across the 110 miles from Denver to Pueblo. He spread his wares out in a vacant mule stable. On the first day of business he took in about $100. Voorhees decided against continuing with the Pueblo venture, so John returned to Denver and resumed his duties as a clerk.

In the spring of 1863, John decided that he was experienced in buying and selling stock and that he would go into the mercantile business for himself. He reasoned that the budding village of Pueblo promised more opportunity and less competition for a general store than did Denver, the metropolis of the area. John took his wages due in merchandise, and with help from Voorhees in arranging credit at several wholesale houses in Denver, John bought a team of mules and a wagon. He loaded his wagon with sale goods and headed south.

John rented store space on the corner of Santa Fe Avenue and First Street. In a letter to his family, he described his place of business as "a board-and-batten building with a board roof overlaid

JOHN A. THATCHER
1836 - 1913

MARGARET THATCHER
1848 - 1922

LENORE THATCHER
1867 - 1890

LILLIAN THATCHER
1870 - 1948

JOHN HENRY THATCHER
1872 - 1928

ALBERT R. THATCHER
1874 - 1877

RAYMOND THATCHER
1885 - 1968

by a foot of dirt sprouting weeds." The countertop consisted of several planks laid on two barrels. The floor space measured ten square feet and the building came with a small iron safe.[5] Later, John moved his inventory and the safe into a larger building at the corner of Santa Fe Avenue and Second Street. It was a time of growth and capitalization on the Front Range of the Rockies and John's business prospered.

Pueblo was the center of a large trade area. The little village of Canon City, along with a few scattered mining camps, lay to the west along the upper Arkansas River, whereas farmers in the valley east of Pueblo grouped in little hamlets. John kept his wagon and hired teamster constantly on the road hauling trade goods, wool, farm produce, pelts and hides, and occasionally a load of ore to and from Denver and settlements up and down the Arkansas River. With an eye to filling a need created by the expanding farming of the area, John and O. H. Perry Baxter combined their assets and financed the builder of a grist mill in Pueblo.

Mahlon D. Thatcher, John's younger brother by three years, was engaged in the family merchandising business in Martinsburg. Excited by the reports in John's letters about the land-office business in retailing and pace of life on the frontier, Mahlon decided to heed his brother's beckoning and head west. He stepped down from a Concord Coach into a March wind swirling dust and dried horse manure along Santa Fe Avenue in 1865.

The brothers immediately went into a partnership and established the firm of Thatcher Bros., Merchants. They moved into a larger building at the corner of Santa Fe and Fourth. Emigration from the East to the gold fields on the slopes of the Rockies had slowed down and only the fading echoes of booming dynamite blasts reverberated between canyon walls. But many newcomers stayed on, and commerce flourished. Puebloites, as well as miners, ranchers, farmers, and transients, passed through the door into the shadowy interior of the Thatchers' store. The front porch, facing Santa Fe Avenue, seemed to be a favorite spot for the spit and whittle crowd to congregate and "chew the fat."

On April 17, 1866, John married Margaret Henry, a teacher and daughter of Judge J. W. Henry of Pueblo. Baxter, John's business associate, married Margaret's sister, Edna, in a double wedding ceremony. It was the social event of the season when two of Pueblo's most esteemed and eligible bachelors took the two lovely young ladies to be their brides. The guests who attended the wedding and reception in the home of the brides' parents ran the gamut from rough-cut men from the countryside with their home-spun wives in tow to townsmen in their Prince Alberts with ladies on their arms wearing the latest silk and brocade fashions from "back East."

In 1866, the merchandise for John and Mahlon's store came to Pueblo from the industrial East on two legs, first by rail to Sheridan, Kansas, the western terminus of the Kansas Pacific Railroad, then by wagons across the prairie to Denver where it was loaded onto the Thatchers' wagons and hauled to Pueblo.

Warring Southern Cheyennes and Arapahoes impeded the westward advance of the railroads. Survey crews carried rifles along with their transits for protection against forays by mounted Indians. On August 6, 1867, near Plum Creek in Nebraska, a party of Indians wrecked a Union Pacific freight train and massacred the crew. It was not until 1870 that gandy dancers laid rails to carry the Kansas Pacific trains into Denver. Two years later, the first Denver and Rio Grande Railroad locomotive, pulling freight cars from Denver on narrow-gauge tracks, blew its steam whistle at a street crossing in Pueblo.

The registered census of Pueblo County in 1872 was 2,323, with 1,067 residing in Pueblo, the county seat. Agriculture and mining were the main industries. Saloons—offering spirits, gambling, and painted ladies—outnumbered all the other retail businesses. Farming was confined primarily to small irrigated plots of land along the valley of the Arkansas River.

Herds of cattle, either trailed in from Texas or spawned by draft animals used by wagoners and from milk stock trailed in behind the wagons of settlers, dotted the plains near permanent water holes and creeks in the Territory of Colorado. John W. Iliff, the first cattle king of the Great Plains, began building his herds by buying cull stock from wagon caravans—the lame, yoke-galled, gaunt, or surplus—and the milk or stock cows from disgruntled miners and settlers who were pulling stakes and going back home. By 1861, Iliff's cattle ranged over some seventy miles along the South Platte River northeast of Denver.

In 1867, Iliff expanded the market for his cattle when he negotiated a contract with the Union Pacific Railroad. He was to supply beef to the railroad's construction crew and its army escort, who were engaged in building grade and laying track into eastern Wyoming. He rode south that fall for a prearranged meeting with Charles Goodnight. They met in the Apishapa Valley northeast of Trinidad, Colorado. A trade was made. Goodnight put $45,000 in his saddle bags, and riding point, led his cowboys and a herd of Texas steers on the long trail to Cheyenne, where he was to deliver the cattle to Iliff.[6] Later events suggest that Goodnight left the drive long enough to ride into Pueblo and stash the $45,000 in the iron safe John and Mahlon had in the corner of their store.

Some Mexican sheepmen headed their herds north from the Mora Valley in the Sangre de Cristo Mountains in the Territory of New Mexico into the open range that lay between the Valley of the Purgatory, north of the Raton Mountains, and the Arkansas River. This move put them closer to the wool warehouses and the market in Denver and farther away from the Navajos who had a propensity for raiding the Mexicans' ranches for stock and slaves. John and Mahlon seized the opportunity and expanded their business by trading in wool and mohair.

On August 1, 1876, Mahlon, thirty-six, married Luna Jordan. She was the daughter of J. O. Jordan, a merchant, banker, and wool buyer in Pueblo.

The transplanted Thatchers from Pennsylvania figured prominently in the history of their adopted state of Colorado and the Great Plains. They were

> the biggest businessmen in Colorado and the best bankers that the State has had. They were a dominant power, but withal most modest and unassuming. For fifty years they walked the streets of Pueblo, where we have never heard a reflection on their moral or business character.[7]

Many of the tracks the Thatcher brothers left over the span of fifty years have been obliterated by the passage of time and flood waters from the Arkansas River that inundated the basement of the First National Bank of Pueblo in 1921 and destroyed their records. Nevertheless, they were motivating men of vision in industry and agriculture. Their actions and influence were interwoven in the development of industry in Colorado and cattle ranching on the Great Plains.

2

DON COLORA'O

Frank G. Bloom was born in 1843 in Blair County in the state of Pennsylvania. As a young man, he worked in Henry Thatcher's store in Martinsburg. Letters from John and Mahlon were shared with Frank, a friend of the family and a suitor of Sarah Catherine Thatcher. Accounts of the brothers' rapidly expanding mercantile business made exciting reading for young Frank. It left him with the feeling that life was passing him by in staid Pennsylvania.

The call to adventure was irresistible to the ambitious young man. In her memoir, *Don Colora'o,* Alberta Iliff Shattuck states that her mother, Sarah, gave reluctant approval to Frank's going west. She figured that a short stay in that "barbarous" country would surely cure him of the "western fever." In the summer of 1865, Frank gave notice to the senior Thatcher that he would be leaving the store at the end of the year to join John and Mahlon.[1]

But an earlier event, a bloody episode in the Indian Wars, changed Frank's plans. On November 29, 1864, Colorado Volunteer militia under the command of Colonel John Chivington, a hellfire-and-brimstone zealot, had assaulted a group of Cheyennes and Arapahoes encamped on the bank of Sand Creek in southeastern Colorado. The Indians, mostly old men, women, and children seeking peace, had turned themselves in to the commander at nearby Fort Lyons and were under his sanction. The predawn attack routed and brutalized the Indians.

Stories of the massacre spread across the Great Plains like a grass fire running before the winter winds, carrying the word to the tribes. Council fires burned late into the night and the unifying message of war drums sounded across the Plains. From the time the tribes left their encampments in the spring of 1865 until winter drove them to their retreats, retaliating warriors waged all-out war. The Sioux and Cheyenne isolated Denver.[2] Travelers

7

beyond the protection of army posts were subject to attack, and stage stations along the trails leading west were burnt to the ground. As a precautionary measure, the army discouraged civilian travel without army escort beyond western Kansas.

It was not until the spring of 1866 that the pursuit of the basic necessities of life overrode revenge, and the thumping of Indian war drums faded. Frank Bloom again gave notice to his employer, packed his truck, borrowed a hundred dollars from his father, and boarded a train heading west.[3] At St. Joseph, Missouri, he hired on as a mule skinner for a wagon train bound for Colorado. A trip of thirty-two days during which the edgy crew kept watchful eyes on the horizon for Indians who might still be wearing war paint.

Those must have been long and tiresome days for Frank, sitting on the plank seat of a jouncing wagon, hunched up against the dust and biting chill of the March winds. There was probably little to break the monotony of the seemingly endless grassland. Frank, no doubt, was fascinated by the bands of curious pronghorns that would cautiously approach the wagons then wheel in unison to speed away, the fluffed white hairs on their rumps flashing in the sunlight. Wagon caravans crossing the prairie would occasionally come across great herds of migrating buffalo flowing unimpeded over the greening prairie like a slow-moving wave of muddy water.

Notwithstanding the Pennsylvanian's appreciation of the beauty of sunrises and sunsets and his interest in the country and the fauna that were so different from home, it was an agonizingly slow trip for a young man who was anxious to join his friends at the end of the line. Step by step, the mules pushed the country behind them. In March of 1867, Frank finally saw the buildings of Pueblo rising out of the brown prairie against the dark background of the Rocky Mountains. John and Mahlon welcomed their friend from home. They expanded the firm of "Thatcher Bros., Merchants" by opening a store in Canon City, thirty miles up the Arkansas River from Pueblo, and putting Frank, their new partner, in charge.

The numbers of wagon caravans traveling westward had been increasing since the signing of the peace treaty reuniting the North and South. Not only were wagons carrying trade goods, but many of the prairie schooners were loaded to the gunwales with the personal belongings of families seeking a new beginning. New settlements began to sprout like mushrooms along the rivers and creeks on the eastern slopes of the Rocky Mountains from the Boseman Trail in Montana to the Santa Fe Trail in Colorado.

Trinidad, tucked among the northern foothills of the Raton Mountains in south-central Colorado, was a major stopping point for wagon caravans traversing the Mountain Branch of the Santa Fe Trail. Felipe Baca and William Hoehne had opened a store there in 1862 that offered some basic supplies. In 1865, the Colorado legislature authorized "The Trinidad and Ratoon [*sic*] Mountain Wagon Road Company." This was Richens "Uncle Dick" Wootton's hacked-out toll road that crossed over the rugged mountain range.

Captains halted their caravans on the banks of the Purgatoire River in the vicinity of Trinidad to ready the loads and animals for the dreaded journey over the mountain. Wagon wheels that had shriveled crossing the dry prairie were soaked in the river to swell the felloes up snug between the iron wheel rims, the spokes, and the axles. Broken spokes, spindles, axles, tongues, and hounds were replaced. Spindles were greased and tie-down ropes on the Osnaburg coverings over the bows and those securing the cargoes were tightened. Horses and draft animals were checked to be sure they did not need to be reshod. Provisions to restock the larder were purchased from local merchants. There would be very little beside pinto beans, chili, and mutton or goat meat available in the Mexican hamlets beyond the pass and north of Santa Fe.

Uncle Dick's toll road was hell on people, wagons, and draft animals. It was a laborious trip of about fifteen miles as the crow flies, but some twenty-seven miles along the bottom or flanks of a narrow, winding, rock-strewn canyon. A cloudburst higher on the mountain could, in the blink of an eye, bring a raging torrent surging around a bend in the canyon. The natural pass rose from six thousand feet elevation on the prairie near Trinidad to almost eight thousand feet at the foot of towering pines on the crest. On the south side, the trail dropped sharply along the winding canyon and over rocky shoulders to the grama grass plains of the Territory of New Mexico. It took three to five days for the caravan to cross, depending on the weather and mishaps along the way. Campsites were wherever the caravan stopped. It might be at midday if there was a breakdown, or in the shank of the afternoon.

Level campgrounds were rare, maybe a meadow on a bench above the canyon. Hay had to be carried along for the animals because forage was sparse on the rocky, forested mountainside.

East-bound caravans carrying wool, hides, or trade goods from Santa Fe or Mexico stopped in Trinidad to readjust loads that might have shifted coming over the mountain, to make needed repairs to wagons, and to tend to draft animals. Supplies needed to supplement the larder and see the wagons and crew across the desolate country between Trinidad and the settlements of Kansas were purchased.

Trinidad in 1867 was described thus:

> A more complete specimen of an American-Mexican town than Trinidad could not be found. It consists of a main street, lined on either side by adobe houses of one story with flat roofs and few rooms. Many of these were "stores" belonging to American traders, and well stocked with goods; two of them were billiard and saloons, and two were boarding-houses—all American innovations. There was no public-house proper, but strong drinks were sold at every one of the establishments and so far as I could make out, at every house in the town. "Liquoring up" seems to be the sole amusement of the inhabitants. It commences before breakfast, goes on all day, and begins again with renewed vigor at sunset.[4]

The growth of Trinidad and its importance as a way station to the mounting traffic along the Santa Fe Trail was not lost on John and Mahlon Thatcher. Frank had shown a flair for trading by turning a nice profit on a twenty-thousand-dollar inventory in the ten months he had managed the store in Canon City.[5] In 1867, the name "Thatcher Bros., Merchants" was changed to "Thatcher & Co." so as to include Frank. The partners agreed that Frank would close the store at Canon City and move the inventory to the more promising location on the west end of Main Street in Trinidad.

Main Street was the only thoroughfare in Trinidad that outclassed a trail; and it was either ankle-deep in mud or shoe-top deep in dust and dried excrement from draft animals and saddle horses, depending on the weather. The village was composed of a few business buildings flanking Main Street and a dozen or so rather mean adobe, log, or board-and-batten dwellings that were scattered at random along the valley.

The partnership paid Frank the princely salary of fifty-four dollars a month for his services as clerk and manager of the store in Trinidad. Frank was a jovial young fellow and a good conversationalist who made friends easily. He would introduce himself to a new customer, and with a twinkle in his eye add that it was *he* who was the *company* in the firm of "Thatcher & Co."[6]

Most of the settlers in and around Trinidad were of Mexican descent, citizens of the Republic of Mexico before the Treaty of Guadalupe-Hidalgo in 1848. According to the treaty's terms, the Mexicans could choose American citizenship, return to Mexico, or stay and be a resident Mexican national. Frank made a special effort to befriend these rather wary people and to learn their language. He was known among his Spanish-speaking friends as "Don Colora'o" [*sic*] for his carrot-red hair and beard.[7]

According to old maps, the historic name of the valley north of the Raton Mountain was *El Valle de Los Animas Perdidas en Purgatorio*. According to the legend that has been passed down over the years and recounted to the author by a homesteader in the area, the name came into being during the time when that area was part of New Spain. A party of soldiers, anxious to hunt for gold, rebelled against their leader and ran away from the main body of *conquistadores* who were exploring the area. The deserters followed an Indian who said he would take them to where they would find gold nuggets as big as gourds lying on the ground. The guide led them into an ambush in a narrow stretch of the valley below the present site of Trinidad where they were killed.

The story is that the soldiers, being Catholic, had perished without absolution. Without the remission of their sins, they were condemned to spend eternity in purgatory. Hence, the English translation of the name given to the valley by the Spaniards: The Valley of the Souls Lost in Purgatory. The early-day mountain men, many of them Frenchmen, couldn't handle this long-winded eulogy. They shortened it to Purgatoire.

A study of numerous letters between Frank Bloom and John Thatcher in the author's files shows that the Thatcher store in Trinidad carried an inventory of general merchandise and sold in either wholesale lots or at retail. Their letterhead advertised dry good, groceries, liquors, hardware, faring implements, queensware, clothing, hats, boots, and shoes. Frank was also the agent for Hess Wagons.

According to correspondence between Joy Poole and Alberta Iliff Shattuck, one of his first orders to St. Louis was for a shipment of wood-burning stoves. He received coal-burning stoves by mistake. During his scouting trips to familiarize himself with the area, he had seen thick outcroppings of coal exposed along the flanks of the hillsides and in the banks of deep washes. As a native of a coal-producing area, Frank had taken more than a casual look at the thick seams of bituminous coal. With the idea of creating a market for the coal stoves he had in inventory, he filed a claim on a section rich in coal and hired a crew to mine it.

At first they brought the coal out of the mine by wheelbarrow. This proved to be too slow to provide for the market that was developing so Frank had the crew use horse-drawn sleds on skids. John and Mahlon were quick to recognize a budding enterprise and filed a claim adjacent to Frank's. The business flourished and wooden rails were laid along the mine shaft to accommodate the coal-pit cars. The coal was sold through the store in Trinidad. A market soon developed in Pueblo, then Denver, and eventually as far away as St. Joseph. Frank was indeed a pioneer in developing one of the largest producing coal fields in the United States. Bituminous coal from the Trinidad area is still being exported to markets around the world.

In July 1869, Frank, accompanied by John Thatcher, boarded a stagecoach in Pueblo. They were on their way to Martinsburg, Pennsylvania. Culminating a long courtship, the last three years by mail, Frank G. Bloom and Sarah C. Thatcher were united in holy matrimony. The happy event was the beginning of a long, loving relationship between Frank and Sarah, or Sallie, which was Frank's term of endearment for her.

In spite of Frank's enthusiastic description to Sallie of life on the frontier, the trip must have been grueling and unsettling for a young, cultured woman from Pennsylvania. The first leg of the journey was in a swaying, hot, sooty railroad coach with stiff-backed, hard seats and a bare minimum of amenities. They shared that part of their return trip to Trinidad with a Lt. Col. George Armstrong Custer and his wife Libbie.[8] Then by stagecoach, with no amenities, they bounced and jarred across the interminable, wind-swept, lonesome prairie.

Looking through the window of the stagecoach as it rolled down Main Street in Trinidad, Sarah Bloom saw her new home.

Mr. and Mrs. Frank Bloom. From the Aultman Studio, 1928. Courtesy Colorado Historical Society.

According to photographs housed in the Bloom Pioneer Museum in Trinidad, the buildings flanking the dusty, littered roadway in Trinidad during those early years were not made of brick and tile with clinging vines and fronted by white picket fences, manicured lawns and rose bushes as they were in Martinsburg, but were of stacked blocks made of dried mud or rough-lumber, many topped with flat roofs covered with sprouting weeds. Swarthy men dressed in rags and naked brown children likely stood beside the road watching the stage carrying the newlyweds as it rolled to a stop. This typically squalid frontier town must have been a cultural shock for the young lady. Her grit and her faith and love for Frank Bloom most certainly saw her through the transition.

3

GOLD NUGGETS AND GOLDEN SPIKES

GOLD! GOLD!! GOLD!!! GOLD!!!!

These words, heralded in gold letters, headlined the September 4, 1858, issue of the *Weekly Press* of Elwood, Kansas. The news was based on rumors of a strike in Cherry Creek in the eastern foothills of the Rocky Mountains. Although stories and rumors likening the find to the California strike at Sutter's Mill proved to be greatly exaggerated, they brought gold fever to a host of migrants who flooded across the plains of eastern Colorado like a tidal wave.

It was this siren's song of gold just waiting in the hills that first brought settlers to the Rocky Mountains. The area, which had been the home of Native Americans for forty thousand years and the stomping ground of a few hardy, solitary, bearded mountain men and explorers for some two decades, had its character and nature forever changed by the invasion of some thirty thousand prospectors and the onslaught of their picks, shovels, and blasting powder.

Most were disappointed in their quest, finding that the glitter in their pans was pyrite. Those who found gold, and there were some fortunate enough to share in the estimated twenty-five million dollars mined from 1858 to 1867, clustered their shanties and dugouts to form settlements. Some settlements endured while others became the ghost towns of western lore. These newcomers to the frontier were not as self-sustaining as were their predecessors—the emigrants who crossed the Bering Strait. The prospectors created a demand for support forces: the army, merchants, bankers, freighters, stockmen, saloonkeepers, prostitutes, lawyers, gamblers, predators, scavengers, and the railroad.

The economy, depressed by the Civil War and its aftermath, was beginning to turn around. Money was becoming available for expansion. In the summer of 1862, President Lincoln signed a bill authorizing government assistance for construction of the long-awaited transcontinental railroads. For every mile of track completed by a railroad through public domain the government would grant ten sections—sixty-four hundred acres—flanking the railroad's right-of-way, each section was checkerboarded with public domain lands.[1] The grants amounted to millions of acres that could be utilized by the railroads, used as collateral, or sold to settlers. During construction, when costs increased in the rougher, more isolated western regions, Congress agreed to double the acreage to be conferred in those areas. Government loan bonds, federal and state subsidies, and the sale of public stock issues helped the railroads finance expansion.

By 1867, the Kansas Pacific Railroad had laid tracks across Kansas and into Colorado. On May 10, 1869, at Promontory, Utah, the last blow by a silver spike maul on the head of a gold-plated rail spike united the Union Pacific Railroad with the Central Pacific Railroad. The populous eastern states were now tied to the gold coast with parallel steel rails. The Kansas Pacific reached Denver in 1870 and then branched north to link up with its parent line, the Union Pacific, at Cheyenne.

Surveyors for the Atchison and Topeka Railroad, followed by construction crews, had headed west out of Topeka, Kansas, in 1868. The Denver and Rio Grande Railroad was incorporated in 1870. Its narrow-gauge rails, thirty-six inches apart instead of the standard 56.5 inches, connected Denver and Trinidad with branch lines serving the gold and silver mines in the Rockies.

The completion of roadbeds to grade and the laying of crossties and rails on trunk lines and branch lines connected the states to the territories. The arterial system promoted cross-country commerce and emigration. The locomotive engineers blew the warning whistles and opened the throttle wide for settlement, mining, ranching, and farming in the Great Plains from the Canadian border to the Rio Grande.

Cattlemen from Texas and those across the wide Missouri River saw the gates opening onto the Great Plains. There, unclaimed except for a few holdout Indians, lay a vast expanse of pristine public domain. It was a world of pasturage where the buffalo grass was sodded thick and big bluestem brushed a mounted man's stirrups. Cool, sparkling waters ran in the rivers and coulees. And there was a market for their beef: hordes of prospectors and miners, blossoming settlements, and contingents of soldiers in scattered forts. Cattlemen had two captive markets—Americans who preferred beef to gamey meats, and Native Americans who were promised meat in the treaties that confined them to reservations.

In an 1868 editorial titled "Future of the Plains" for the *Cheyenne Leader*, Nathan A. Baker remarked:

> That a future of the greatest importance is in store for the western plains, no one who has traveled over and lived upon them for any considerable length of time can doubt . . . this country will make stock raising a specialty since it is one of the most profitable branches of industry which it is possible for civilized man to engage in.[2]

TRAIL DRIVE

In 1866, Nelson Story sewed ten thousand dollars in paper currency under the lining of his clothes and with two companions rode horseback southeast from the territory of Montana toward Texas. His aim was to purchase longhorn steers and trail them back to Virginia City, and sell beef to miners and settlers. He bought six hundred head in the Fort Worth area. With a crew of twenty-six Texas cowboys, who wanted to see what was on the other side of the hill, he pointed the herd northwest for thirteen hundred miles.[1]

Texas cattlemen returning home after soldiering for the Confederacy found their ranges overgrazed and overpopulated with cattle. During five years of war, the prolific herds of longhorns had multiplied. Cows, bulls, and mossy-horned steers ran wild in the brakes of eastern Texas among the thorny mesquites, which grew taller than a man on horseback. The supply of stocker cattle and beef far exceeded the local demand. It was said that a good baying hound with a sharp nose was worth more than a dozen good-aged cows.

Also in 1866, a young Texan named Charles Goodnight decided to hunt for greener pastures for his cattle. He had heard stories of the grassland to the northwest where great herds of buffalo roamed, multiplied, and fattened on nutritious grasses. With a binding handshake, a partnership was formed between Goodnight and Oliver Loving, an older man who had made a nice profit on a herd of longhorn steers he trailed to Denver six years earlier.[2] On a spring day, somewhere south of Weatherford, Texas, Goodnight led a crew and a herd of two thousand cows, bulls, and steers west toward the dark horizon.[3]

To protect the herd against a probable attack by warring Comanches and Kiowas, Goodnight figured to go south of the Western Cross Timbers, the tribes' hunting ground. He and Loving thought that the cowboys might be two months or more in the saddle driving the herd a thousand miles, or more, on what they hoped would be a successful and profitable venture. They reasoned if they couldn't market the cattle, they could at least find greener pastures.

With young Goodnight on the point and the crew spread out on the swings and the senior Loving with the drags, the herd was lined out along the trace of the old Butterfield Overland Stage route. They moved the cattle westward, twelve to fifteen miles a day, for about 120 miles as the crow flies, then turned west by south. They forded the North Concho River near the present-day site of San Angelo, made good time the next day, and at dusk circled the cattle on a bed ground along the north side of the Middle Concho. At first light Goodnight pointed the herd west along the Middle Concho.[4]

They had moved the herd south of the country over which the Indians held dominion—some 220 miles without incident. The Llano Estacado, or Staked Plains, lay ahead. It was a vast expanse of sameness without the grace of shade from a tree. Deep, sandy loam laid down by an ancient sea rolled across a vast plateau like gentle swells on an ocean. A carpet of grama, buffalo, and waving bunchgrasses stretched from horizon to horizon. There were no permanent waterings, only seeps, wet-weather creeks, and shallow lakes after a rain squall, that soon dried up.

> Imagine yourself . . . standing in a plain to which your eye can see no bounds. Not a tree, nor a bush, not a shrub, nor a tall weed lifts its head above the barren grandeur of the desert.
>
> Albert Pike, "Narrative of a Journey in the Prairie," 1834

Goodnight's crew watered out the cattle, filled their canteens and water barrels out of the bubbling spring that fed the Middle Concho, and then pushed on.

A cow in the heat of summer will drink eight to ten gallons of water a day to maintain body condition. The cowmen faced a formidable task in pushing the herd eighty miles to reach the waters of the Pecos River. On the second night the cattle were too thirsty to bed down. Goodnight put all the hands except the cook and horse wrangler on night guard circling and singing to hold the restless, milling stock on the bed ground. At first light they headed the bawling cattle westward. It was a three-day, three-night fight against thirst, dust, and balky cattle for the men and a walk in the spreading shadow of death for horses and cattle down that last long stretch through the valley of Centralia Draw.

They crossed the Pecos River with its alkaline water and bogs of quicksand at Horsehead Crossing. Then they pointed the herd north along the Pecos River, an arid country claimed by hostile Apaches, rattlesnakes, and scorpions. Goodnight found it to be a sorry country, hardly fit for grazing. He called it "the graveyard of the cowman's hopes."[5] The coral-hued ground between the clumps of salt grass and cacti was daubed with white crystalline alkali. The infrequent runoffs drained into the saline, muddy waters of the meandering Pecos from flanking arroyos, their sides cut steep and deep by the eroding waters, making the cattle drive slow and difficult. Almost a month to the day since they left Weatherford, they arrived at Fort Sumner, Territory of New Mexico.

Lady luck smiled on the two gambling cowmen. The garrison at Fort Sumner was responsible for providing sustenance for 8,577 Navajos and 485 Apaches confined on nearby Bosque Redondo Reservation. The supply contractor at the fort offered to buy all the marketable-aged beeves in the Goodnight and Loving herd. A price of eight cents per pound, on the hoof, was agreed on and sealed with a handshake.

The sale did not include the breeding cattle. The partners decided that Goodnight would take most of the proceeds from the sale, some twelve thousand dollars in gold, and with three of the crew go back home, put together another bunch of cattle, and trail it west in the tracks of the first herd. Loving and the rest of the crew would push the stock cattle on north into Colorado.[6]

The consensus among cattlemen, commission men from the eastern markets, and drovers was that the prime rangelands of the bighorn country of Wyoming and the Black Hills of western Dakota Territory would be available for pasture cattle only when the Indian problem was resolved.

Buffalo hunters were making a killing, in two senses of the word, on the northern plains. The government was promoting the use of buffalo hides by eastern tanneries as a source of leather, and the army was protecting the hunters against retaliation by the Indians. This clandestine action was aimed at reducing and scattering the Plains Indians' cloven-hoofed provisioner, the strategy being that privation would bring the Indians to the peace councils. There they would be promised reservations and sustenance, sign the treaty, and surrender their firearms.

The defeat of the Seventh Cavalry and the death of the flamboyant George Armstrong Custer and 224 troopers in 1876 was victory for the Sioux. But it was the impetus that roused the public to demand that the Indians be brought to their knees and that an all-out war be waged to bring the Indian problem to a conclusion.

Reinforced army units were put in the western field. A concerted effort by battle-toughened veterans fragmented and scattered the tribes and seized control of the Black Hills and the Bighorn Basin, areas coveted by gold prospectors and cattlemen. By 1877, all but a few of the Plains Indians were beaten and starved into submission. The tribes were gathered and herded onto government-assigned reservations. They were placed under the supervision of Indian agents who would administer the peace and see to the distribution of the promised dole.

Railroads and eastern publications touted stories of great profits to be realized by grazing cattle on the western frontier. *Rocky Mountain Husbandman* quoting the Buffalo *Livestock Journal,* November 25, 1875 recounted: "Cotton was once crowned King, but grass is now. If grass is King, the Rocky Mountain region is its throne, and fortunate indeed are those who possess it."[7] An initial investment of twenty-three dollars for a cow and her steer calf could be parlayed into from forty-five dollars to sixty dollars in four years when the calf had matured to a grass-fat animal ready for market.[8] This thirty percent return made sense to the drawing-room investor who had read Walter Baron von Richthofen's book, *Cattle Raising on the Plains of North America.* The 1885 publication asserted that in ten years a herd of one hundred cows would propagate to 2,856 head. Compounding this intriguing thought was James S. Brisbin's *The Beef Bonanza, or How to Get Rich on the Plains.* Brisbin showed that $250,000 borrowed at ten percent and invested in cattle would yield $810,000 in five years.

The penciled expenses figured by the cattle-baron-to-be included cows, bulls, a crew of twenty-dollar-a-month cowboys, some horses, a chuck wagon with a cook, and a registered brand. For each additional thousand cows owned, the per-head cost would be proportionately reduced and the profits per cow thereby increased. The frosting on the cake, the cream on the pail, was the thick-sodded, nutritious grasses waving over the millions of acres of public domain. The aspiring cattleman merely needed to purchase a few acres around watering places to give him range rights to a limitless spread of adjoining prairie.

Investors looked on this as a capital idea. Speculators from abroad financed some of the largest operations: Prairie Cattle Company, Ltd.; Matador Land and Cattle Company, Ltd.; Swan Land and Cattle Company; XIT of Texas; and Maxwell Land and Cattle Company, Ltd. By 1883, a few major investors had their brands burnt on the hides of over half of the twenty-three million cattle grazing on the midcontinental prairie.

FREE GRASS

According to the Preemption Act of 1841, a settler squatting on public domain could purchase 160 acres of the land he claimed by preemptor's rights for $1.25 per acre. Then, as a further inducement to settle the Great Plains, President Lincoln signed into law the Homestead Act of 1862. This granted to a head of a family or adult male a clear title to 160 acres of public land. The only limiting provisos were that the petitioner had not borne arms against the government, would pay the small filing fee, agreed to maintain a residence on the property for five consecutive years, and farmed an appreciable portion of the quarter section. The South had seceded and civil war was imminent. By the terms of the Homestead Act, soldiers for the Confederacy would not be eligible for land grants.

> The public domain, or public lands—the property of the nation, and subject to legislative control and disposition by Congress alone—is the area known as public lands acquired by treaty, capture, cession by States, conquest, or other acquisitions and purchase, and lying and being in the States and Territories....
>
> The several existing laws for the sale and disposition of public domain permit entries and location by individuals, associations, and corporations.
>
> The theory of the settlement laws is that an individual, if he be not already the owner of 320 acres of land, can purchase 160 acres under the pre-emption act after six months' settlement, occupation and improvement, and can acquire 160 acres under the homestead act by residence, improvement, and cultivation for a term of five years,... or can purchase at the end of six months by commutation.
>
> Under the several settlement and occupancy laws, however, a person can legally acquire 1,120 acres of the public domain.
>
> The existing laws recognize several classes of lands:
>
> Mineral.—"In all cases 'lands valuable for minerals' shall be reserved from sale, except as otherwise expressly directed by law."

Timber and stone.—Lands valuable chiefly for timber and stone, unfit for cultivation.

Saline. —Salt springs.

Town-site lands.—Any unoccupied public lands.

Desert. —Lands which will not, without irrigation, produce an agricultural crop.

Coal lands. —Lands containing coal.

And all others as agricultural. Agricultural lands can be taken from 40 to 160 acres under the pre-emption, homestead, and timber-culture acts, or purchased at public sale or private entry.

Of agricultural public lands there are two classes: the one class at $1.25 per acre, . . . and the other at $2.50 per acre. . . . The latter class consists of tracts embraced within the alternate sections of land reserved to the United States in acts of Congress making grants within prescribed limits of the lines of railroads, or other works of internal improvements, to aid in the construction thereof, such reserved sections being double in price. . . .[1]

To facilitate and encourage development of the West the army built a series of garrisons along the travel routes. Mounted troopers campaigned against the Plains Indians and offered protection to settlers and transient caravans. Migration from the East accelerated.

As the Civil War escalated, the Union recalled troops from the western frontier to reinforce armies engaged in combat with the South. The Indians were quick to take advantage of the situation and renewed, with a vengeance, their attacks against settlers, wagon trains of commerce, and would-be settlers. Development of the Great Plains slowed to a crawl during the war years as the Indians, only hindered by solitary guns of the settlers and an undermanned army, shoved back at those who would invade and lay claim to their ancestral lands.

As the 1860s came to a close the tremors of an economy suffering a hangover from the binge of war were calming. The dawn of the 1870s saw expansion of the railroads and a return of troopers to arm garrisons on the western frontier, thereby promoting and facilitating a resumption of emigration. The railroad provided a more efficient and economical way to market raw materials and products of the Great Plains in the populous and industrial East.

The bison on the Great Plains faced extermination from the army of hide hunters. The herds near the Republican and Arkansas Rivers were nothing more than a memory. All that remained of these great herds was a stench and a rick of bones. The hunters moved their stands south to the Texas herd. The steady booming of their long-range rifles sounded across the Comanche and Kiowa country as far south as the Concho River drainage.

Most of the Indian leaders had signed peace treaties and their people were herded onto reservations. A few freedom-fighters still struggled against the surrender of their ancestors' burial ground. In the desert region of the southwest the wily, tough Apaches still held development of the territories of New Mexico and Arizona in suspension by terror. The barriers were down and a movement to settle the unexploited lands of the Great Plains picked up another head of steam. Land speculators and the railroads, with their millions of acres of grant lands, ballyhooed the pot of gold awaiting settlers at the western end of the rainbow which arched the Great Plains.

Masses of people seeking a plot of land to call their own gathered together their worldly goods and trekked after the sun. The first waves dug their plows into the rich black soil under the blue skies of western Iowa. Followers leapfrogged to the thick-sodded buffalo grass country of Kansas, Nebraska, and the eastern Dakota Territory. They soon found that they were farming in a far different environment than they were accustomed to back home. The average rainfall in the short-grass country was less than in the Appalachian Mountains and the moisture-sapping, eroding winds never seemed to lay.

The homesteaders' struggle to get roots embedded in the soil was disheartening. The Economic Panic of 1873 and a severe drought were followed in 1874 by a living apocalypse of locust which stripped the crops to stubble, denuded the pastures, shredded drying laundry hanging on the line, and chewed up tack and harness.

Hopes and dreams, topsoil from plowed fields, and the sanity of many of the isolated women in their windowless dugouts or sod-brick hovels were all swept away with the howling winds. Many settlers either sold their rights to their homesteads for whatever they could get or walked off and left them. The westerly advancement of sod-busters was halted in its tracks or at least slowed to a crawl along the more fertile river valleys.

Rolling hills blanketed with nutritious grasses and herbs that could condition cows to cycle, spark the libido of bulls, and put

27

leaf lard and pones on fattening steers lay waiting. Millions of acres of idle pasturage stretched from Canada to Mexico and from the eastern foothills of the Rockies to the hundredth meridian. Nature backed up this sea of grass with abundant water coursing along the rivers and creeks and bubbling springs that fed the ponds. The plant and the raw materials necessary for large-scale ranching were at hand.

There was a growing demand for slaughter cattle in the East, where the plow had turned under much of the grasslands, and for beef from the army at its reactivated western garrisons. The government was contracting for all classes of cattle as a source of meat for annuities promised the subjugated Indians. The market for slaughter cattle on the frontier was increasing with the growth of the population, and aspiring cattlemen and speculators provided an active market for breeding stock.

The cattle industry on the Great Plains had grown from infancy in the early 1860s to a burgeoning state at the breaking of the 1870s. The base plan was simple. By taking advantage of the Preemption or Homestead Acts, legally or otherwise, or by outright purchase, a cowman could patent 160 acres or more, controlling permanent water rights. The laws were "used only to be abused." Cowboys filed on land that was later sold to their employer for a nominal sum within twenty-four hours after title had been secured. At other times men were paid to preempt locations, or fictitious names for nonexistent people were used for filing. With this base for his livestock, he was a rancher.

Most cattle watered every day but some would graze further out and only return to water every couple of days. A cow suckling a calf required about twenty pounds of forage a day. A conservative cattleman would estimate the number of acres needed to pasture a cow and her calf for a year, dividing 640 acres by the average annual rainfall, and stock his country accordingly. For example, an area receiving an average of twenty inches during the growing season would call for thirty-two acres per animal unit per year. But for the majority of the ranchers grazing public lands, greed overrode prudence. Overgrazing the range soon became the norm on the prairie east of the Continental Divide as the cattle cycle ballooned and cattlemen figured more was better.

A man could drive a herd of cattle to his deeded acreage, acquaint the animals with the watering place, and turn them loose.

The herd could then range out at will. The cattleman thereby laid claim to range rights on public domain that lay adjacent to his deeded land to the extent that it was utilized by his cattle. A claim to range rights was based on a first-come priority and was backed by the unwritten laws of the frontier and the code of the cattlemen. Infringement on another's established range was an unacceptable breach of the code and was most often settled by a Colt .45 or Winchester .30.30.

In 1868, Frank Bloom attended a meeting of cattlemen of southeastern Colorado who gathered at Fort Garland to hammer out regulations governing the grazing of livestock on regional public lands. It was agreed that cattle should be herded from the middle of May to the first of November, but that they might run "without let or hinderance" the rest of the year. In November 1871, they met again and formed a stock raisers' association, planned the arrangements of roundups and the recording of brands, condemned the driving of cattle except by owners, condemned the running of inferior bulls on open range, and advised sheepmen to "pay due and proper regard to existing claims of cattlemen to ranges already occupied by them." The group recommended fencing cultivated fields against livestock and the branding of calves by the time they were six months old.

Other actions taken by the association at that meeting were the establishment of an acceptable code of ethics for participating cattlemen, the affirmation of the unwritten laws of the range designed for the common good of the members, and the sanctity of the property rights of all. Henry C. Thatcher, who had followed his older brothers, John and Mahlon, to Colorado, was engaged to serve as the association's lawyer.

John A. Thatcher. Courtesy Pueblo Library District.

Mahlon Thatcher. Courtesy Pueblo Library District.

THE DAWNING OF A NEW DAY

In the 1870s, the eastern slopes of the Rocky Mountains were bustling with activity. There were still occasional raids by the Utes and the Arapahoes, and sometimes forays into the foothills by Comanches, the horsemen of the plains. But by and large the fragmenting of the tribes by the U.S. Army and the swelling population of armed settlers were deterrents to Indian resistance.

The swarm of prospectors who had combed the mountains over the last decade for precious minerals had laid the groundwork for a prosperous mining industry. Solitary, bearded men leading pack burros still prowled the dark canyons, picked at granite outcrops, shoveled pay dirt into their sluice boxes, and panned the gravel beds of mountain streams. But miners, some with families, were banding together to establish nascent settlements.

Railroads were engaged in a headlong race, grading their road-beds westward, vying to complete rail service from the East to the West Coast with branch lines to growing communities in the Great Plains. The motivations for urgency were the federal, state, and territorial land grants being awarded for each mile of completed track, and the anticipated freight revenues.

This transition moved the Great Plains from the epoch of a raw, undeveloped frontier toward an era of civilized growth. The dark, bloody days were nearing to a close and a family-oriented society was being wrested out of the wilderness by resolute women-folk and a few civic-minded men. There was a world of prime agricultural land and a bonanza of raw materials to be exploited. A wide and lucrative variety of commercial ventures awaited entrepreneurs. These inducements were drawing people and industry from the states. The old log, adobe, or board-and-batten buildings, often with sod roofs and dirt floors, that housed the early-day businesses were being replaced by masonry buildings, some several stories high.

The various enterprises of John and Mahlon Thatcher were riding the crest of the wave. They seemed to have the Midas touch. The brothers shared a congenial relationship, strengthened by a mutual respect for their individual abilities. Each shouldered his part of the responsibilities with effectiveness. Apparently, John was the public relations member of the partnership. He had an innate liking for his fellow man and a natural rapport with people. His friendly but reserved personality, coupled with an unpretentious demeanor, attracted people and instilled a feeling of confidence. Mahlon was the more reserved of the two. He made certain that their business ventures were structured efficiently and that they maintained a firm grip on internal control. He had a keen understanding of the intricacies of the world of finance and investments. Both men were extremely discerning when it came to appraising the character of people. As expanding businessmen and promoters of the growth of Pueblo and Colorado, they were recognized as responsible captains of industry.

Their mercantile store in Pueblo and affiliates in outlying villages were doing a land-office business. The iron safe that John had acquired along with his first store was the stimulus for a new endeavor for Thatcher Bros., Merchants. People in the territory, whether local residents or transients, took advantage of the box to safeguard their valuables. Neither John nor Mahlon bothered to give receipts. They simply sealed the currency or gold dust in a canvas sack, wrote the owner's name on it and stuffed it in the safe. This grew to be such a popular service that the brothers ordered a two-ton strongbox from St. Joseph, Missouri.

Men with currency tucked in money belts or gold dust or coins in their saddlebags found it prudent to cache their valuables. Word got around that one could bank on the Thatchers' integrity.

The fireproof safe was a squat, black box that measured forty-four inches high, twenty-four inches wide and sixteen inches deep. It was made of quarter-inch boiler iron and had a combination lock.[1] The box became the cornerstone on which the fortunes of men and a good part of the advancement of the western frontier was built. The popularity of people banking in their safe led the Thatchers to offer the added financial service of buying and selling drafts drawn on established banking institutions. On January 26, 1871, the brothers announced that they had expanded their operations to include Thatcher Bros., Bankers. Their bank was located in a newly erected, two-story, "fireproof" building on the corner of Santa Fe Avenue and Fourth Street, across from their mercantile store.

In February 1871, Mahlon boarded a Southern Overland stagecoach in Pueblo for Washington, D.C. to apply for a charter for the First National Bank of Pueblo. Reference banks were Kountze Brothers Bankers of New York, Union Banking Company of Philadelphia, and Colorado National Bank of Denver. Incorporators were John A. Thatcher, Mahlon D. Thatcher, O. H. Perry Baxter, Charles Goodnight, Benjamin D. Fields, Henry C. Thatcher, Henry W. Cresswell, and Jefferson Raynolds. John and Mahlon were the only stockholders. Elected officers of the bank were John Thatcher, president; Fields, vice president; Mahlon Thatcher, cashier; and Raynolds, assistant cashier. The First National Bank of Pueblo acquired the assets of Thatcher Bros., Bankers and opened its doors at Fourth Street and Santa Fe Avenue with a capital stock of fifty thousand dollars. At the end of the year it listed resources of $176,004.30.[2]

John, Mahlon, Baxter, and Cresswell were partners in a farming and cattle operation east of Pueblo. Goodnight was the Texas cattleman who had followed Oliver Loving to Colorado and established a cattle ranch. Henry Thatcher had moved to Colorado in 1866 and was an attorney in Denver. He later became chief justice of the Colorado Supreme Court.

The incorporators of The First National Bank of Pueblo were men in their thirties, the prime of life. They were ambitious men of action and shrewd businessmen whose vistas were not bounded by horizons nor shadowed by clouds of doubt; visionaries who would, as a group and as individuals, leave their mark on the history of the Great Plains.

In notes taken during an interview on September 16, 1927, at the Goodnight home in Goodnight, Texas, J. Evetts Haley, biographer and author of *Charles Goodnight Cowman and Plainsman*, recorded that in Charles Goodnight's opinion, among those who should receive credit for their work in bringing law and order to southern Colorado were Hank Cresswell, John, Mahlon, and Henry Thatcher, and Perry Baxter.

The directors of the First National Bank of Pueblo could soon see that the growing community could support another bank. The population of Pueblo was almost doubling with the passing of

Frank Bloom and James Sutherland. Courtesy Pecos
Valley Collection, Chaves County Historical Museum.

each year. More and more of the native sod in the Arkansas Valley
was being turned under by the plowman and the surrounding
range was stocked with growing herds of cattle and sheep. The
loan officers of the First National Bank of Pueblo were swamped
with requests by businessmen needing funds for inventory and
growth, by farmers for financial aid in producing crops and by
ranchers applying for loans to finance herd expansion.

In September 1873, the Stock Growers' Bank opened its doors
for business on the corner of Third Street and Santa Fe Avenue.
The directors of the new institution were Perry Baxter, Henry
Cresswell, Charles Goodnight, and Jefferson Raynolds. Financing came from the First National Bank of Pueblo and from
the directors.[3]

THE ASSOCIATES

As a young man in Indiana, Oliver Hazard Perry Baxter was a journeyman blacksmith. In 1852, he moved to Moline, Illinois where he worked for the John Deere farm implement company. He continued his migration westward in 1856, moving to Nebraska City, Nebraska. In 1858, the twenty-three-year-old Baxter and three companions joined the stampede of prospectors to the Territory of Colorado. They walked from Nebraska City to the Pike's Peak region where gold had been found in Cripple Creek canyon.

Baxter prospected and mined for several years along the Front Range. As a commissioner appointed by Governor Gilpin, Baxter helped organize Pueblo County in 1862. He was a member of the Colorado Volunteers and was with Colonel Chivington in the attack on the Cheyenne and Arapahoe encampment on Sand Creek in the winter of 1864. During his hundred-day enlistment with the militia, he was elected to represent Pueblo County in the Territorial legislature. In 1867, Baxter and John Thatcher set up a partnership and financed a grist mill at the southwest corner of Fifth and Main in Pueblo. When the owner was unable to repay the loan, the partners foreclosed and Baxter took over the management duties.[1]

Baxter was a civic-minded man who took an active part in state and county politics. He was an inveterate venture capitalist with wide interests in banking, ranching, railroading, and real estate development. The friendship and business association with the Thatcher brothers continued until he died in 1910.

The old cowboy testimonial: "He's a man to ride the river with," fit Henry W. Cresswell. He was born in England in 1830 and came to the Great Plains, by way of Canada, in about 1859. Perry Baxter and "Hank" Cresswell met in the gold fields. They hit it off and spent several years prospecting, until they decided to

Roundup in the Arkansas Valley of Colorado, circa 1885.

turn their hand to farming. They located acreage suitable for irrigation near the confluence of the St. Charles and Arkansas Rivers east of Pueblo. In 1862, they raised a bumper crop of barley, beans, and corn which they sold to the government at Fort Lyons for a profit. In 1863, the partnership expanded its agricultural operation to include the raising of cattle. They purchased 142 head of stocker cattle for seventeen dollars a head.

Through Baxter, Cresswell became acquainted with John and Mahlon Thatcher. In 1870, the Thatchers provided the financing enabling Henry to start a hundred-cow dairy operation near Pueblo.[2] The brothers kept a close eye on the operation and were impressed with the broad-shouldered, personable bachelor's industriousness, his knowledge about cattle and farming, and his straightforwardness. It may well have been a reaction to Cresswell's contagious enthusiasm and optimism over the future of raising cattle that put the Thatchers, and eventually Frank Bloom, into the livestock industry.

> Hank Cresswell was . . . stout, well built, with a rosy, weather beaten face and a quick active step. . . . Withall, there was a naive reserve, some would call it shyness, it took some time to break into the secrecy of his soul. . . . During our visits he related many a story of frontier life, of prospecting in Colorado. One day he was riding up a canyon in

Southern Colorado. It was about noon that he was looking for a shady spot to camp when two shots rang out from behind a boulder, two bullets whistled past his head and two Indians stood up behind the pile of rock. They came out boldly and began loading their rifles. Cresswell had just gotten a repeating Winchester. His pony had stood fire like an old military mount, as probably he was. The Winchester came out of the scabbard and one Indian promptly dropped. The other with a ghoulish laugh and a gleam of revenge in his eyes evidently thought he could feel the white man's scalp already in his hand, when Hank pumped another shell into the barrel and just as the redskin was placing the cap on the nipple of his gun another shot broke the silence. He jumped six feet into the air and turned a somersault, never gave a kick. . . .

Cresswell was one of the most modest of men. He scarcely ever used the personal pronoun, and yet he was brave as a lion.[3]

In 1868, Charles Goodnight established a Colorado ranch along Apishapa Creek northeast of Trinidad. It was to serve as a swing station for his trail herds coming north from Texas through the southeast corner of Colorado. In the winter of 1869 he relocated his ranch headquarters to a sheltered valley of the Arkansas River east of the Front Range and about five miles west of Pueblo. His cattle grazed along the drainage basins of the Arkansas and St. Charles Rivers. Goodnight considered this open range to be good cow country. It was thickly sodded, primarily with grama and buffalo grasses. These were strong, short-stemmed, deep-rooted, nutritious perennials that thrived in the deep topsoil over limestone bedrock. There were branch canyons and stands of scrub oak, cedar, and piñon trees for protection from winter winds. The growing population in neighboring Pueblo was an added enticement to the cattleman and it was only about a ten-day trail drive to an active cattle market in Denver, capital of the territory.

Goodnight turned his cattle loose to range with the deer and antelope in the foothills and on the prairie that stretched eastward from the Rockies. By the unwritten law of range rights he laid claim to all the land that was marked by the hoof prints of cattle wearing his brand (P A T), a vast spread that ran from north of the Arkansas River to south of the St. Charles.

In 1874, the H. W. Cresswell and Company was organized. Their brand, the Bar CC (ᑕᑕ), was registered in the livestock brand book of the territory of Colorado. The company was a partnership. The investors were John and Mahlon Thatcher, Perry Baxter, and Henry Cresswell. Cresswell was the man in the field.

They ranged their cattle on public lands that lay between Hardscrabble Creek and Charlie Goodnight's range rights along the St. Charles River drainage.

Frank Bloom was an astute employer and businessman. As a brother-in-law, employee, and respected business partner of John and Mahlon Thatcher, he was the resident manager of the Thatcher & Co. mercantile store in Trinidad. He oversaw the mining and merchandising of coal for himself and the Thatchers out of their operations northwest of Trinidad. Always a civic-minded person, he volunteered his time and efforts in promoting the growth of Trinidad. His commitments ran the gamut from treasurer of the Bible Society to treasurer for the Las Animas Railway and Telegraph Company.

In 1871, the Thatcher brothers, through their bank, had received a quit-claim deed from Johnson Barritt to the Hole-In-The-Rock Ranch located some thirty-five miles northeast of Trinidad. They registered a Circle Diamond (◇) on the left ribs with the livestock board, and with Frank overseeing the operation, they began to stock the ranch. This patented land, surrounded by public domain, gave the Thatchers exclusive possession of a dependable watering for their livestock at the head of Timpas Creek. This natural cistern of sweet water had for years been a welcome oasis for travelers along the Mountain Branch of the Santa Fe Trail. The rock barn and dugout Barritt had built on Timpas Creek became ranch headquarters for the hired cowboys who were looking after the Thatchers' growing herd of cattle and horses.

Charlie Goodnight and his P A T cattle neighbored the H. W. Cresswell and Co. range. The partners of the Bar CC and the transplanted Texan could see that the frontier east of the Rockies was coming of age. Their "irons were hot," as the cowman would say. It was time for them to tally and brand the increase of their herds.

A breeding herd of cattle running on the open range under favorable conditions could double its population in three years. Grass and water were free for the taking and cowboys worked from "can 'til can't" for fifteen to twenty dollars a month. On the Denver market in 1871, yearling steers were selling for $10, two-year-olds for $15, long-ages for $25 to $30, dry cows at $18 to $20, and cows with calves from $21 to $23. On the Chicago market, slaughter steers were bringing $4.50 a hundred on the hoof.[4]

In that same year the Thatcher Brothers & Co. stores were selling a hundred pounds of flour for $4.50, coffee at 24¢ per pound, a hundred pounds of sugar for $18.50, a hundred pounds of pinto beans for $4.00, a yard of flannel at 65¢ and a yard of jean at 90¢.[5]

Henry Cresswell, Cresswell Land and Cattle Company.

BLACK FRIDAY

Goodnight and Loving broke trail. News of the good market in the north spread across the rangelands of Texas like a prairie fire running before a March wind. It was not long before drovers were shaking the bushes on the hills and riding out the draws of central Texas gathering cattle to take advantage of this better market. The shouts of cowpunchers and the popping of bull whips resounded across Texas ranches as cattle were gathered and thrown on the trail northbound for Colorado and southern Wyoming. The men riding drag tied their neck rags over their noses against the clouds of dust and swarms of horn flies that hovered over the passage of thousands of cattle. It was estimated that by 1869 approximately a million head of cattle were grazing on the eastern plains of Colorado.

The cattle that grazed in the wake of the bison in the late 1860s were a crossbreed of Texas longhorns that had come up the trail, cattle used as draft animals on the wagons from the East, and the milch stock trailed in by immigrants. Their hides resembled the coat of many colors that Isaac, in the Old Testament, gave his son Jacob. There were paints, calicos, line-backed or bonnet-faced browns and blacks, solid whites, and other varied combinations. The longhorns coming from Texas traced their ancestry back to the Spanish cattle from Andalucia. Their lance-tipped horns, straight or curved, were massive and could measure six feet or more from tip to tip. They provided the longhorn cow with protection against predators that would prey on her calf, giving that breed an advantage in raising a calf to weaning age. Some of the cattle had short horns, a genetic carry-over from Hereford, Jersey, or Durham cattle that traced their heritage to England or Europe.

For the Thatchers, acquiring cattle and horses in the early 1870s to run on their Hole-In-The-Rock property and adjoining public domain was almost incidental. Merchandising, banking, and mining were their primary concerns. They would, on occasion, take livestock as payment on accounts. Frank, on a collection trip down the Purgatoire Valley in 1869, had not only accepted a herd of cattle from John Anderson against his indebtedness at the store in Trinidad but stayed the night at the Anderson home and attended a wedding the next day. As years passed, the number of Circle Diamond cattle and horses under Frank's management grew. The Thatcher ranching operation began to expand on public domain between the headwaters of the Timpas and the Apishapa Creeks in Las Animas County.

On Black Friday, September 19, 1873, the effects of the collapse of the financial empire of Jay Cooke, the East Coast mogul, swept across the Great Plains. Without the underwriting support of Cooke the construction of the Northern Pacific Railroad stopped. The Railroad's ambitious plans for hyping the sale of acreages adjacent to the right-of-way of a transcontinental line left many immigrants holding an empty sack: 160 acres in the desolate Northern Plains miles from any settlement with no transportation other than a wagon or afoot.

The collapse of Jay Cooke's bank reached across the continent, touching many segments of the economy. Nervous bankers called in loans and tightened purse strings. The stagnation was most devastating in the West where growth and development was dependent on the availability of capital. The air was let out of the ballooning livestock industry.

John and Mahlon had invested heavily in real estate in Pueblo. They deeded the land over to the city rather than pay taxes. Charles Goodnight said of his investments in Pueblo:

> The panic ... wiped me off the face of the earth. I had loaned $6,000 on a half block of ground on which was the only brick building in the town; I also owned the opera house and all the vacant buildings in the town, [which] would just about pay the taxes in 1873.[1]

He later sold the half block for two thousand dollars, and according to Mr. Haley, Goodnight "was mighty glad to get it."[2]

The November 12, 1874 issue of the *Star Journal Chieftain* ran a legal notice under the dateline: Pueblo, October 19, 1874:

Having found it necessary to employ our available funds in our own private enterprises, we, the undersigned have this day disposed of our respective interests in the Stock Growers' Bank of Pueblo . . . .

O. H. P. Baxter
Henry W. Cresswell
Charles Goodnight

The same issue of the newspaper carried the following advertisements:

FIRST NATIONAL BANK OF PUEBLO
Capital and Surplus $130,000.
Does a General Banking Business.
THATCHER BROS.
WHOLESALE
and
RETAIL
Dealers in
DRY GOODS
CLOTHING,
GROCERIES,
NOTIONS,
Boots and Shoes,
and
EVERYTHING.
Santa Fe Avenue   Pueblo, Colorado.

TRAILING TO TEXAS

The build-up of cattle, sheep, and horses grazing on the prairie of southeastern Colorado in the mid-1870s began to crowd the range that Charles Goodnight claimed. Overstocking, compounded by several years of drought, had stressed the grasses and depleted the available forage. Adding to Goodnight's worries were heavy financial losses incurred during the shock waves that followed the Panic of 1873. He was heavily indebted by speculative investments that were carrying compounding interest rates of from one and a half to two percent per month.[1] Would-be neighbors, range-jumpers, were trying to elbow in on his pastures. He knew he was running too many cheap, thin cattle on depleted range. The annual reports of the Chicago Board of Trade for the years 1864 to 1890 shows cattle prices bottomed out in 1875, and weather records for the years 1873 to 1876 log a three-year dry spell on the Front Range in Colorado.

Riding over his drought-blighted range in the fall of 1874, the pasture grasses chewed off to the crown, the bottoms of the natural lakes seamed with cracks in the gray soil, and his cattle as poor as town dogs, Goodnight's thoughts likely turned homeward. In his youth, as a Texas ranger during the Civil War, he had ridden over some of the country west of Fort Belknap at the southern tip of the Cross Timbers and east of the Llano Estacado. Very likely he recalled the world of waving prairie grasses blanketing open country and rolling hills that supported thousands of buffalo. He compared the vivid memory with the bare range that lay before him.

It was time to make a change. The country west of Fort Belknap was still the land of buffalo, Comanches, and Kiowas; open range waiting for a cowman to put his brand on it. In the spring of 1875, Goodnight turned his back on his Rock Canyon Ranch in Colorado. His cowboys gathered the P A T cattle from the range he had claimed by preemption, range rights, and might. While he rode off to Texas for another trail herd to sell, a trusted man was in charge of moving his cattle and remuda down the long trail back to Texas; leaving behind all that he and his wife Mary Ann had called home since 1869.

The cowboys eased the herd down the broad Arkansas Valley. The character of the cattle was some different from that of the first herd the boys had driven up the trail in 1866. Goodnight, with a desire to improve his stock, had introduced some Shorthorn herd bulls, adding a touch of refinement to the offspring of the original longhorn cows. He summered the cattle that year along Two Butte Creek below Las Animas in the southeast corner of Colorado. That fall he moved south across the line into eastern New Mexico to winter the herd of some sixteen hundred head south of the Canadian River near the Texas line.[2]

The following spring he drifted his cattle alongside the Canadian into the Texas Panhandle; then drew reins to let the herd summer along the Alamocitos, a little tributary of the Canadian.[3] When the grasses began to cure out in the late fall, the crew gathered the cattle and trailed them south by southeast toward the Llano Estacado, that uncharted, unsettled, seemingly limitless plateau that stretched from the Red River country in West Texas to the valley of the Pecos River in New Mexico. In the same year that Goodnight was gambling his all on a venture beyond the horizon, the *Texas Rural Register and Immigrants' Handbook* advised that it was unlikely that the Staked Plains could ever be adapted to human habitation.

Scouting ahead of the slow-moving herd, Goodnight met a Mexican mustanger who told him about a canyon he had seen once when he was chasing wild horses on the Llano Estacado.[4] The old fellow said it was *muy barranca,* a very deep canyon, that there was good water and grass along the canyon floor, and that there were many *búfalo.* The mustanger thought it was the place the *comancheros* called *El Cañon del Palo Duro,* the rendezvous where they went to trade with the Comanches. The way the man described it, Goodnight decided to go and take a look-see.

He hired the Mexican to guide him to this *cañon grande.* They prowled the plains for days, making dry camps or following game trails to find scarce waterings. They used buffalo chips for tinder on this treeless plain. Finally, one day, when the Mexican guide was about to admit that he was lost, they came to the brink of a canyon where the sod ended abruptly and they were standing on the edge of a chasm that rent the crust of the earth.

As they rode eastward the separation between the rim rocks widened and the canyon deepened. As the stockman looked down on the backs of hundreds of buffalo grazing along the canyon bottom beside a meandering stream he knew he had found his Garden of Eden.[5] The almost vertical canyon walls would help act as a barrier against cattle straying and also protect them against the killing northers that were known to sweep across these plains.

Closing out his interests in Colorado took Goodnight along the backtrail to Pueblo on several occasions. There was the business of gathering and disposing of his cattle that were missed on the initial roundup and those P A T strays that Cresswell and his cowboys brought in as they prowled the Bar CC country. During late-night conversations, Cresswell began to compare the range the Bar CC cattle were running on and the future of raising cattle in Colorado with Goodnight's description of the plains of West Texas.

The partners in H. W. Cresswell and Company, like Goodnight before them, were beginning to feel the pressure of stockmen crowding in on their range rights on public lands. And coming over the eastern horizon, like a cloud of dust in the vanguard of a storm, was word of a newly patented invention: wire for fencing that had spaced, pointed barbs that would turn back or confine livestock. It was touted as being "lighter than air, stronger than whiskey, and cheaper than dirt."[6] Cattlemen agreed that it would bring radical changes to the grazing habits of livestock running on open range. Compounding the partners' concerns was their overstocked, depleted Bar CC range south and east of Pueblo.

Goodnight's Rock Canyon Ranch neighbored the Thatchers' ranges along the Arkansas River, and he was certain to be acquainted with his neighbors, Pueblo's bankers. It can be conjectured, in view of the following facts, that Mahlon, Perry, and Henry sat in John's office in the First National Bank of Pueblo and listened to Goodnight. They agreed that the rangeland around Pueblo was overcrowded, with trail herds still coming in from south Texas. Goodnight, in his conservative, blunt way, painted a promising picture for the future in running cattle on the state lands in the Texas Panhandle. It was decided that Cresswell should go back with Goodnight, look at the country, and report back to the Thatchers.

Apparently, Cresswell made a wide circle. After returning to Pueblo in the fall of 1876 he gave a favorable report to his partners. He described the thick turf and the tassels of matured bluestem that existed on the plains of the Panhandle at that time. He probably reported that water for cattle was not as plentiful as it was in the country around Pueblo, as indeed it was not; but that vast stretches of Texas state land along the Canadian River and its tributaries lay unclaimed, as history confirms. As Cresswell may have pointed out, the market for beef would not be what it was in Colorado, but from the Canadian River country, in the upper Panhandle, their cattle could be within trailing distance along the trail Jessie Chisholm had laid out from his trading post on the Canadian River to the railroad stockyards in Dodge City.

In the spring of 1877, Henry Cresswell and the cowboys of the Bar CC rounded up the "dogies," loaded the wagons, and lined out for the wilderness of the Texas Panhandle. The winds of time have covered his trail. But he may have followed the waters of the Arkansas River, or maybe the Cimarron, to somewhere near the Kansas border; then, depending on the availability of waterings, turned south. They probably crossed the western tip of Indian Territory in what is now the Oklahoma Panhandle and headed for Mustang Creek, which heads in the rugged hills of the northwest corner of the Texas Panhandle. This would have led the Bar CC outfit to the Canadian River.

Cresswell's first camp was on a little, willow-flanked tributary of the south fork of the Canadian River. He dubbed it Home Creek, and like most early settlers on the frontier his first dwelling was a dugout cut into the bank of the creek.[7]

Cresswell's camp was not far from the buffalo hunters' outpost called Adobe Walls. Forty years earlier, William Bent had established a post on Adobe Walls Creek to trade with the Comanches and Kiowas; and in 1864, at this spot, Colonel Kit Carson and a contingent of men fought a strong force of Indians from forenoon until nightfall. In 1874, a gathering of the tribes had attacked twenty-eight men and a woman barricaded behind the mud-plastered, palisaded walls of the buildings at Adobe Walls. The young Comanche warrior, Quanah, son of an Anglo woman, Cynthia Parker, and the Comanche chief, Peta Nocona, had a horse shot from under him as he led one of the attacks. Bones of the horses that had been killed in the fierce battle still littered the area when Cresswell arrived.

John Thatcher received a letter which touched on the problems Cresswell was encountering in establishing a ranch for the partnership in the wilds of Texas:

Adobe Walls Texas

Jan 18th 1878

Friend John

I rec'd two letters from you a few days ago by the way of Elliot. I am now camped about 25 miles east of Adobe Walls on the Canadian about 35 miles north of Elliot have most of the cattle in here doing well and am still gathering it has been a hell of a job to gather the herd in the winter especially as the fire split the herd in two and left them forty miles apart on a new range I think however that I will not lose many except what the Kansas grangers & Indians have killed Every damed Indian in the country had been here for the last 6 weeks Pawnees Arapahoes Cheyennes Kiowas Comanches Apaches & Utes have had to watch the saddle horses pretty close have not lost any yet but think I will lose some of the Bronchos as I have had so much to do I could not watch all and concluded they were the least valuable and stuck to the cattle & saddle stock. Think now I will keep the cattle here may however take the beef up on Beaver next spring. Have not had any snow here but about two weeks cold rain which was just as bad. It looks like spring here now—raining and warm. You spoke of buying more cattle how much per cent do you propose to charge the concern if we buy and when would the money be available and how much. Write and let me know the particulars in your next as I may see a chance here in the spring and if so could avail myself of it. I am going to see the Commanding officer at Elliot and try to get pay for what the Indians have killed. They did it on the sly in the Canons and deny killing any have found some and know in reason they have killed more trailed them to their camps but they had hid the meat and have no direct proof except in one case they said they would pay for it but sliped off without doing so. Some white men stole some of their horses the other day The Indians and some soldiers went after them found them They let the soldiers into their camp then took their horses arms etc and kicked their sterns and told the soldiers to go back which they did Indians and all I will write again soon.

Yours truly

H. W. Cresswell

Charles Goodnight. Courtesy Nita Stewart Haley Memorial Library, Midland, Texas.

BEEF BONANZA

Change came to the Great Plains during the 1870s. Settlements were becoming villages and villages towns. The provincial municipalities of Denver and Pueblo were acknowledged by eastern cousins as attaining the status of cities. Transcontinental rail service, the resurgence of land-hungry immigrants—many from Europe—the introduction of the steel plow, barbed wire, the windmill, and Hereford bulls were changing the features of the land and the character of agriculture. Politically, Colorado moved from a territorial frontier to statehood in August 1876. By 1880, its population had increased by some 154,000.

Cattle branded with a Circle Diamond on the right ribs and hip and ear-marked with a swallow fork in the left ear and a hole in the right ear (℗—♋) were grazing on the prairie of Colorado north of the Purgatoire River from Trinidad to La Junta and westward to the St. Charles River. The Thatchers had acquired spotty acreages of patented lands in the Las Animas, Bent, Huerfano, and Pueblo counties, which gave them access to thousands of acres of adjoining public lands.

The pressures of their varied and growing business enterprises tended to confine John and Mahlon to their desks. Frank Bloom was spending more and more time overseeing their expanding cattle operation and correspondingly less time looking after the store in Trinidad. In the spring of 1875, Thatcher & Co. elected to close the store. That year, members of the newly organized Las Animas County Stock Growers' Association elected Frank Bloom to head the organization.[1]

A letter dated December 2, 1878, from Frank Bloom to John A. Thatcher, Esq., Martinsburg, Pennsylvania, stated in part:

> We have in the herds about 3,678 head including calves. We branded
> 848 up to the time I saw the book last. Will have a few more. I worked

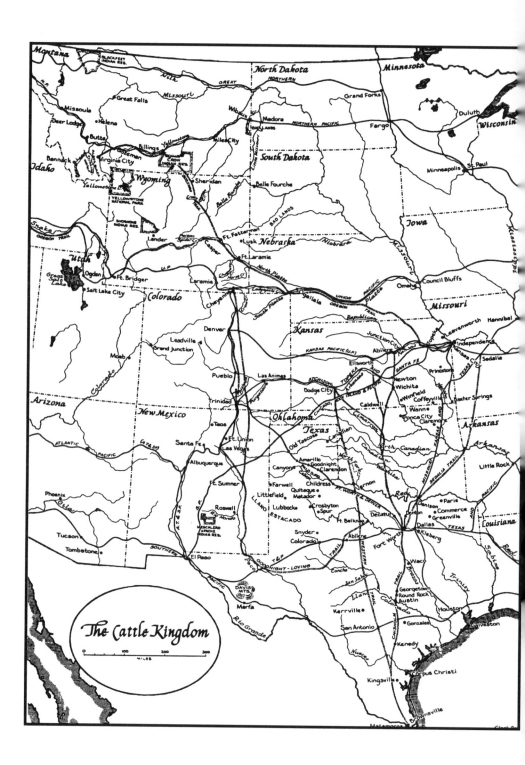

The Cattle Kingdom

0   100   200   300
        MILES

up a good many cows this summer. I am very pleased with the Herefords. I must have some more of them.

The sea of grass, abundant flowing waters, and a growing market for beef that Goodnight saw in southern Colorado in 1867 had, for a decade, been the pot of gold at the end of the trail for Texas stockmen. The heaviest concentration of cattle and sheep was on the prairie east of the foothills of the Rockies; south of the South Platte River and north of the Raton Mountains. It was estimated in 1869 that there was something near two million sheep on the open range in Colorado. North of the Platte, in the lands of the Northern Cheyennes, Sioux, and Crow, stockmen in the late 1860s were still subject to losing their herds and cowboys their scalps. The country south of the east-west range of mountains that lay between Colorado and New Mexico had not yet caught stockmen's attention because it was known not to have the rainfall nor the many rivers that were to be found north of the Ratons. It was an enigmatic land where carpetbaggers dominated politics, unfriendly Mexicans the rural villages, and Apaches the southern region.

The news of quick, sizable profits being turned by cattle ranchers spread from the banks and hotel lobbies of Denver and Cheyenne to the financial districts of New York and Chicago. The wireless carried the message to financial institutions and private clubs in London and Edinburgh. To blue-blooded swashbucklers, tales about the romance and stirring adventures of wild, mounted cowboys fighting off hostile Indians as they drove great herds of cattle across hundreds of miles of virgin prairie, pumped the adrenalin. And to the more mature, less adventuresome, the stories about the easy money to be made from investments in ranches and livestock was as heartening after the stagnation that followed the Panic of 1873 as is the coming of spring after a long, dreary winter.

Articles like the one that appeared in the August 1878 issue of *Harper's Magazine* whetted the appetite of investors:

Q. What profits may be expected in the stock business?

A. The following may be pronounced a fair and reasonable commercial estimate, and it is put forward with only the remark that while the figures apply to circumstances as they are now, and there are

chances and contingencies and possible disasters attending money-
making adventures of all kinds, the margin here is so large that after
making all allowances which caution may suggest, one has still the
promise of great results.

We will suppose an individual or a firm to have found a ranch to
suit him or them in Southern Colorado, and to have bought it. The
cost is hard to fix; but one of 10,000 acres, in complete order, could not
stand in at more than $50,000.

A herd of 4,000 good cows could be bought at $18 each or $72,000.
And 80 good short-horn and Hereford bulls at an average of $50 each,
or $4,000. Making a total investment of $126,000.

By careful buying in the spring one should get 70 per cent of calves
with the cows, or say 2,800 calves. Of these, on the average, one-half, or
1,400, will be heifer calves.

At the end of the first year affairs should stand as follows:

The 1,400 heifer calves will be yearlings, and worth $14,000. There
will be 1,400 yearling steers, worth $10 each = $14,000. Total =
$28,000.

With a herd of this size expenses may be put at not more than
$5,000. And for contingencies, sundries, and ordinary losses it is safe
to take 4 per cent on capitol invested in stock, say, on $76,000 =
$3,040. Total expenses = $8,040.

Profit at end of first year = $19,960.

At the end of the second year the 1,400 heifers are two years old, and
worth $5 more apiece, or say $7,000. And of the 2,800 (70 per cent of
4,000) new yearling calves, an average of one-half, or 1,400, will be
heifers, and worth $10 each, or $14,000; 1,400 two-year old steers are
worth an additional $6 each, or $8,400. And the 1,400 new yearlings
are worth $10 each, or $14,000. Total = $43,400.

Deduct expenses: $5,000. And 4 per cent on capitol invested in
stock ($76,000 + $19,960 = $95,960), or $3,838 = $8,838. Profit for the
second year: $34,562.

At the end of the third year the original 1,400 heifers are three years
old, and worth an additional $3 per head, or $4,200. The yearling heifers of last year are two years old, and worth an additional $5 each, or
$7,000. There are 1,400 yearlings from the original stock, worth
$14,000. And of the offspring of the three-year-olds (70 per cent of
1400 = 980) one-half, are heifers, and worth $4,900. The original 1,400
steers are three years old, and worth an additional $10 each, or $14,000.
The 1,400 steer calves of last year are two years old, and worth an additional $6 each, or $8,400. And there are 1,400 yearlings, offspring of the
original stock, and 490 offspring of new three-year-olds, in all, 1,890 at
$10 each = $18,900. Total = $71,400.

Deduct expenses on 5,400 cows, say $6,050. And 4 per cent on
($95,960 + $34,562) $130,522 = $5,221. Total expenses = $11,271.

Profits at the end of the third year $60,129. Total net profit for three
years: $114,651.[2]

By the beginning of the 1880s, the excitement of speculative investment in the Great Plains rivaled that of the gold and silver rush in 1859, only in 1880 the enticing commodity was free pasturage on public domain and a ready market for beef. The war whoops of the Indians were no more than dying echoes. Locomotives were pulling freight cars transporting cattle to and from the railroad stockyards that dotted their transcontinental lines and refrigerated cars were carrying swinging carcasses from slaughterhouses in the midcontinent to markets back East. The economic shock waves from the quake of 1873 had subsided, and investment capital was becoming more available.

The scent of money wafting from investments in cattle ranching on the Great Plains quickened the pulse of capitalists, here and abroad. Investors, many who knew no more about the bovine critter than that the species was herbivorous, the source of beef and milk, and reputedly self-sustaining, were eager to become stockholders in a cattle venture. Their collateral, and source of anticipated lucrative dividends, was generally the seller's "book count," a range tally taken by hired cowboys of cattle running loose on some distant, vast prairie in the wilds of the West, or as viewed by English and Scotch investors, in the outback of the Colony.

Eastern journals carried enticing stories about "cattle barons" riding horseback or in surreys over their western kingdoms and how they were reaping great profits from cattle ranches. The book, *Beef Bonanzas; or How to get Rich on the Plains,* made the rounds of the corporate board rooms and private clubs. The author, General James S. Brisbin, was not a cattleman but a cavalryman who had spent twenty-three years of his life serving on western army posts.[3] He probably didn't know which end of a cow got up first. (The rear end first on a bovine and the front of an equine.) Nevertheless, Brisbin professed an expertise in the volatile business of raising cattle.

Goodnight and Cresswell were not the only cattlemen who were feeling the pressure of crowded and depleted ranges in Colorado. From 1870 through 1879, over three million head of cattle had been trailed north out of Texas. Added to this weight on the ranges were the natural increase of breeding stock and the great flocks of sheep grazing on public lands. Testimony taken by the

Public Lands Commission in 1879 confirmed that ranges were indeed overstocked and the grass was playing out.

Indian summer came to an abrupt end when, in early December of 1878, a paralyzing blizzard roared down the Front Range from Wyoming to the northern plains of New Mexico. Two storms followed, each on the heels of its predecessor, covering the prairie east of the Rockies with three to four feet of snow. Bitter cold set in after the storms and the surface of the snow crusted over. Horses could paw through the snow to reach the buried grass, but the only pasturage for cattle and sheep was where the strong winds blew the snow off the ridges, and the brush that was available along rivers and in the coulees. On the open ranges that provided only scant protection from the freezing winds and where livestock were expected to rustle for themselves, the chinook in February came too late. Thousands of dead cattle and sheep dotted the prairie.

Like the Bloom Cattle Company, the JJ Ranch in Las Animas County in southeastern Colorado suffered great losses during the winter of 1878. The owners were aware of the growing interest investors were showing in ranching, and they could read the signs of impending overgrazing of the open range. In 1881, they decided that it was time to cash in their chips. They sold the JJ brand, the range rights they claimed on 2.25 million acres of public domain along the Purgatoire, some three hundred horses, and fifty-five hundred cattle to the Prairie Cattle Company, Ltd., Scotch investors, for $625,000 in sterling.[4]

LOOKING SOUTH

In 1880, the region that lay south of the forested crest of Raton Mountain, the rocky rampart between Colorado and New Mexico, was a land of physical extremes. The topography ranged from alpine mountains where the annual moisture could exceed forty inches to desert plains that might get ten inches in a good year. It was a part of Nueva España from 1591 to 1821, then a province of the Republica de México until 1848, and now a territory of the United States. It has a history of being peopled by conflicting and varied cultures: Indian tribes, Spanish-Mexicans, and latter-day Anglo-Americans.

The Rocky Mountains pierce New Mexico from the north. Spaniards, who saw the rising sun turn the snow-capped crests of the mountains blood-red, named the southern tip of the Rockies the Sangre de Cristos. Convulsions of the earth's crust during the Pre-Cambrian period left volcanic, mountainous islands and brushy, basalt-capped mesas standing alone in the surrounding plains.

In 1852, Col. Edwin Voss Sumner submitted an official report to Washington recommending that the government "return New Mexico to the Mexicans and Indians."[1] In January of 1874, Gen. William Tecumseh Sherman, testifying before a senate committee, declared that "ownership of the Territory of New Mexico is not worth the cost of defense."[2] To Texas cattlemen, who had followed the Goodnight-Loving Trail along the Pecos River country from Horsehead Crossing to Fort Sumner, it was an arid land of sparse rainfall, coarse bunch grass, and alkaline water.

During the 1860s and 1870s, when mass migration from the East was settling up the Western frontier and the great trail drives were moving herds of longhorn cattle onto the public domain of the Great Plains, the land west of Texas and south of Colorado was viewed by would-be settlers and stockmen as a

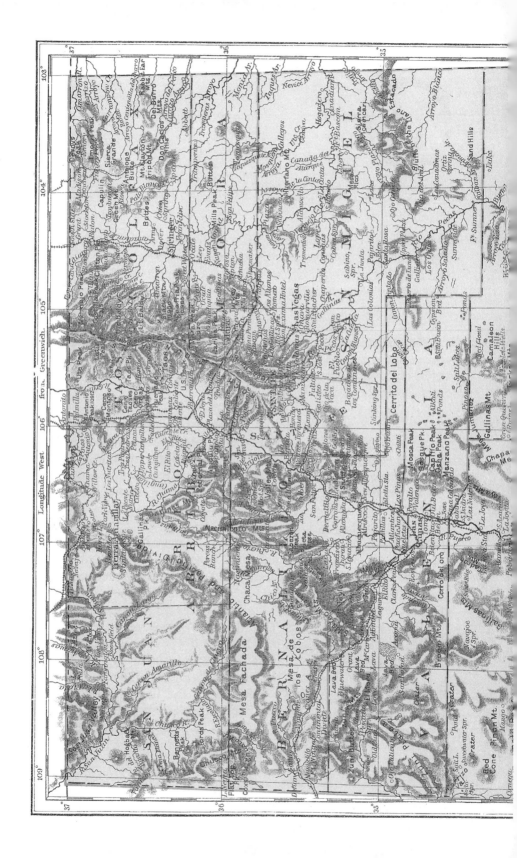

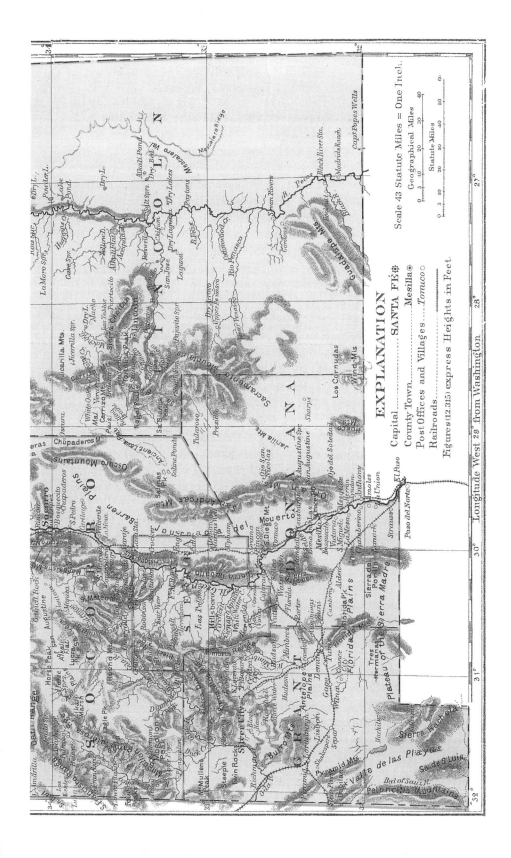

EXPLANATION

Capital...............SANTA FÉ ⊛
County Town.................Mesilla ◉
Post Offices and Villages....Tornuco ○
Railroads........................
Figures (12,315) express Heights in Feet.

Scale 43 Statute Miles = One Inch.

Geographical Miles

Statute Miles

Longitude West 29° from Washington.

desolate, uncivilized land inhabited by Mexicans and hostile Indians. The main commerce for outsiders in the frontier was trade at the end of the Santa Fe Trail.

Excluded to the stockman hunting for grass on public lands were some thirty million acres in New Mexico and southern Colorado. These were tracts of land granted by the king of Spain and by the president of México during their respective reigns. Most of these grants were located along the broad valleys of the Rio Grande and on the eastern slope of the Sangre de Cristos. By the terms of international law, the United States, in the Treaty of Guadalupe-Hidalgo, agreed to give citizenship to the grantees and to issue a quit-claim title to grants validated by the U.S. government.

Lincoln County, as it was originally surveyed, took in most of the southwest corner of New Mexico. In relief, the county was a land of contrast. From the grassy plains west of the Llano Estacado and the harsh, spiny Chihuahuan Desert, the land tilted upwards to forested slopes with rollicking rivers that coursed through the valleys from the crests of the White Mountains and the Sacramento Range toward the Pecos River.

The region was spawned in violence. In the beginning, volcanoes spewed forth a long tongue of molten, wavy, black magmas (the Valley of Fires) that snaked down a valley between new-born upthrusts (Tularosa Basin). The first of the surviving *Homo sapiens* to inhabit the area were the Stone-Age people who called themselves *T'nde*. The exploring Spaniards called them Apache, a name they adopted from the Zuni Indians' word *Ápachu*, meaning enemy people.[3]

John Chisum, a trail-driving cattleman from Texas, found a good market for longhorns west of the Staked Plains in the Pecos Valley. Like Goodnight and Loving, Chisum found the soldiers at Fort Sumner hungry for beef and the confined, starving Indians at Bosque Redondo hungry for any part of the "white-eyes'" buffalo but the inedible horns and the hooves. He established a ranch headquarters and built a four-room house and a stable on the east bank of the Pecos River near Fort Sumner. Chisum trailed in and sold thousands of Texas cattle to the contractors who supplied the soldiers and Indians at Fort Sumner and Bosque Redondo, and to Goodnight, who drove them on to the northern markets. Chisum ranged his cattle along the Pecos Valley from the Fort Sumner area some 150 miles, grazing the arid country flanking the Pecos River out as far as a cow would walk from water. By 1875, he had moved his headquarters to a flowing, sweet-water spring called South Spring River located near the confluence of the Rio Hondo and the Pecos near the present town of Roswell. His reputed sixty thousand head of cattle were branded from shoulder to hip on the left side with the long rail and ear marked with what the cowboys called the jingle-bob.

As the rest of the nation settled into a state of relative peace after the Civil War, inflammatory conditions in Lincoln were causing violence, beyond the occasional raid by Apaches. There were age-old conflicts over racial and cultural differences. Troubles were also brought on by the arrival of hardened outlaws fleeing Texas lawmen, and the preference by most men to carry six-shooters on their hips. Over-indulgence in cheap whiskey and brandy available in the numerous saloons compounded the situation.

For Texas cattlemen moving westward to take advantage of the frontier that lay beyond the Pecos River, an ingrained animosity toward the "greasers" dated from the bloody fighting between the soldiers of the Republic of México and those who had fought for freedom under the Lone Star banner. The native Mexicans, farmers with their little irrigated plots and small herds of livestock and whose ancestors had colonized the area, were antagonized by the Tejanos' arrogance, their attention to their wives and daughters, and their rough and rowdy manners.

Lincoln County was a powder keg with a sputtering, short fuse. The very tone of daily life was charged with tension. Property and wealth were up for grabs: range and timber rights on the vast public domain; lucrative beef contracts to those who supplied the military and Indians reservations; any property held by Mexicans; and gold and silver prospecting rights. Livestock could be rustled by anyone who could alter a brand with a running iron. Those who were powerful could manipulate anything.[4]

Law-abiding men of Lincoln County found it not only prudent but often a matter of self-preservation to pack a pistol during waking hours and to sleep with it at night.

In 1882, Roswell, trying to bring law and order, imposed a fifteen-dollar fine and court costs against anyone found carrying a six-shooter within the village limits. In his book, *Robert Casey and*

Frank Bloom. Courtesy Colorado Historical Society.

the Ranch on the Rio Hondo, author James D. Shinkle quotes the old-timer, C. D. Bonney: "In most any crowd in Lincoln, until about 1885, there would be more guns on the men than there would be men." The mixture of hard liquor, guns, racial animosity, and men who lived by their own code made for a potent and volatile brew.

WEST OF THE PECOS

Many of those who settled on the Great Plains came as drifters after the California gold frenzy had subsided. Jack Sutherland and his stepson, William E. Anderson, disillusioned over not finding the mother lode, drove a band of brood mares and stallions marked with the Scissors brand from the West Coast to the Texas Panhandle sometime around 1877. Sutherland bought the remnant of the old Adobe Walls trading post and turned to ranching. As neighbors to Cresswell, Sutherland and Anderson came to the attention of the Thatchers and Bloom. Perhaps on Cresswell's recommendation they financed their operation with loans from the First National Bank of Pueblo. According to the story, William T. Thatcher told biographer J. Evetts Haley on August 25, 1932:

> Sometime about 1877, Mr. Bloom bought 25 mares from a herd from California in the Scissors brand. He raised all his horses from these original 25 and three or four years ago he told me he had sold over 11,000 horses and about the same time he had 18,000 horses in Canada.
>
> When he bought the ranch near Roswell, he sent all his geldings down there from his headquarters ranch at Thatcher, Colorado, and all of his mares to Montana. This was about 25 years ago. Later he moved them into Canada.[1]

Anderson and Sutherland had no doubt driven their livestock across southern New Mexico on the way to Texas. Apparently Anderson was impressed by the country that sloped eastward from the crest of the Sacramento Mountains to the Pecos River. Water flowed from springs and melting snow packs in recesses of the mountains through lush valleys to the Pecos. Rocky hillsides flanked the valleys and rose to tablelands matted with luxuriant grasses. The Rio Hondo carried abundant water for farming. Except for a few villages and preempting settlers along the valleys, and the Mescalero Apache Reservation of some four hundred

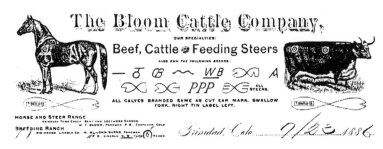

Bloom Cattle Company letterhead.

thousand acres in the Sacramento Mountains, the nineteen million acres in Lincoln County were public domain.

It is likely that William Anderson described his journey across southern New Mexico and the advantages of raising cattle in Lincoln County to the Thatcher brothers. According to a letter Frank Bloom wrote to the Thatchers they were already making loans in northern New Mexico:

Trinidad, Colorado

February 28, 1876

Mr. John A. Thatcher
West Las Animas, Colorado

Dear Sir:

I got back from Mexico last evening. Could not find Watrous. [Author's note: The village of Watrous, New Mexico, was named after Samuel Watrous, an early-day trader and rancher.] He has not been home and the general impression is that he is not coming back soon. I gave him five days to get home after crossing the Apishapa and it is only 50 or 55 miles. You can do nothing by suing in Colfax County as there seems to be no law or officers to execute a suit at Cimarron.

Apparently John and Mahlon were impressed enough with Anderson and his description of the frontier west of the Pecos, to offer financial backing under a partnership agreement for an exploratory venture in Lincoln County.

In the spring of 1882, Anderson boarded a southbound Santa Fe passenger train in Pueblo. He detrained in Las Vegas, which

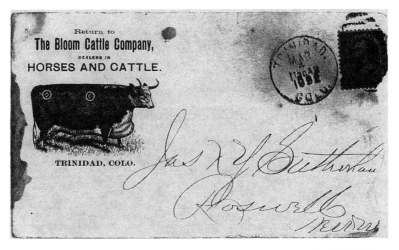

Bloom Cattle Company envelope. Letter from Frank Bloom, Trinidad, Colorado, March 1888 to James Sutherland, Roswell, Territory of New Mexico.

had grown from a Mexican hamlet to a bustling town of commerce during the heyday of the Santa Fe Trail. At a wagon yard near the depot he loaded his bedroll, valise, camp, and horseman's gear into a buckboard and headed southeast across the prairie. Clipping along behind a team of mules over the wagon road that followed the meandering Pecos River, he probably spent the first night at Anton Chico, a village that was old when Brigadier-General Stephen Watts Kearny stood on a rooftop of a building on the plaza in Las Vegas in 1846 claiming New Mexico for the United States. The trace Anderson probably followed led to Fort Sumner before heading south along a well-traveled road that led through John Chisum's range to Roswell, a growing settlement for the past ten years.

An occasional report from a six-shooter or a carbine fired in anger still echoed off the slopes of White Mountain, and the stealing of cattle was a full-time occupation for outlaws in Lincoln County when William Anderson headed his buggy west out of Roswell and up the Hondo Valley. The view of the snow-capped White Mountain and the watershed country west of the Pecos seemed to offer the possibility of better range country than did the alkaline country in the Pecos Valley. Anderson may have followed the Rio Hondo across the tilted plains and into the narrow valley that wound through the sugar-loaf hills formed by alternating strata of limestone and alluvium. Two days out of Roswell, Anderson would have come to a fork in the river. Branching northwest, along a valley that led to the village of Lincoln, ran Bonito Creek. Coming to the junction from the west was the Rio Ruidoso (Noisy Water River).

The further Anderson rode through the foothills the more obvious it became that this was indeed cow country. Grama grasses covered the hillsides and tabletops. The river bottoms were lined with deciduous trees and lush with pasturage. To the north of Bonito Canyon was the backbone of the Capitan Mountains, lying west to east. The southern climate and the protection of the Capitans from killing northers made for an ideal year-round range for livestock.

No doubt, Anderson was excited by the heavily-sodded pastureland that he saw, the long growing season here in the south, the abundance of river water for irrigation, and a climate ideally suited for breeding cattle, sheep, and horses. Spurred on by the fact that the country was sparsely settled, he hurried back to Pueblo.

On October 6, 1882, The Anderson Cattle Company was incorporated in Pueblo. The Certification of Incorporation was filed in New Mexico, November 4, 1882.

Anderson returned to the Hondo Valley in the spring of 1883. He filed for a homestead on a quarter section with water rights on the Hondo River. With a focus on the vast public domain, prime grazing land that flanked the river on both sides, Anderson began an aggressive campaign to acquire river frontage with water rights out of the Hondo. Purchases were financed by the Thatchers and were made in the name of the Anderson Cattle Company.

Anderson and Jack Sutherland had sold their rights to the Adobe Walls Ranch along with their cattle and horses to the Hansford Land and Cattle Company, Ltd., headquartered in Edinburgh, Scotland, and under the management of James M. Coburn of Kansas City. Sutherland let the Scissors brand go with the horses. William Anderson brought his brand, the Diamond A ($\bigwedge$), with him to Lincoln County.

Lincoln County, prior to 1889, contained thirty thousand square miles; all the grazing country of southeastern New Mexico. The central parts of the county were well watered by running streams, the principal of which was the Rio Hondo, draining the

Sierra Blanca and Capitan Mountains. Beside this are the Feliz (or Felix), Ruidoso, Bonito, Eagle, Peñasco, and Nogal Creeks.

> For pasturage and as a stock country, Lincoln County has few equals. The varied kinds of grasses are most abundant and nutritious feed for stock, summer and winter, while the mountains and foothills furnish the best of natural protection in winter. . . .
>
> The profit on cattle here is at least fifty cents monthly per head from the time they are calved, while the profit on sheep is not less than fifty per cent. Circumstances are so favorable to stock-raising in Lincoln County that prudent managers think two per cent is a very liberal estimate of loss from all causes while the cattle and sheep are on the range. . . .
>
> The bountiful and lasting pasture; the excellent climate, where snow scarcely ever covers the grass, combine to make this country an ideal stock region.[2]

THE BLOOM CATTLE COMPANY

As the 1870s drew to a close, John, Mahlon, and Frank could read the message written in the clouds of dust that were sweeping across the overgrazed, droughty rangelands of southeastern Colorado. The range rights of preemptory cattlemen were being encroached upon by newcomers who were filing on homesteads. The three partners were taking steps to build a firm base for the expansion of their livestock operations by acquiring patented lands, especially parcels that had surface water available and adjoined public lands.

The following was recorded in the general land office:

THE UNITED STATES OF AMERICA

To all to whom these presents shall come, Greetings:

Homestead Certificate No. 396. WHEREAS There has been deposited in the General Land Office of the United States a Certificate of the Register of the Land Office in Pueblo, Colorado, whereby it appears that pursuant to the Act of Congress approved 20th May, 1862: "To secure Homesteads to actual Settlers on the Public Domain," . . . the claim of Frank G. Bloom has been established and duly consummated, in conformity to law, for the west half of . . . in township twenty nine south, or range sixty west of the Sixth Principal Meridian in Colorado containing one hundred and fifty five acres and eighty nine hundredths of an acre according to the official Plat of the Survey of the said Land. . . .

You know ye, That there is, therefore, granted by the United States unto the said Frank G. Bloom the tract of land above described: To have and to hold the said tract of Land, with the appurtenances thereof, unto the said Frank G. Bloom and to his heirs and assigns forever; subject to. . . , as provided by law.

In testimony thereof, I, James A. Garfield, President of the United States of America, have caused these letters to be made.

Given under my hand, at the City of Washington, the ninth of April, in the year of our Lord one thousand eight hundred and eighty one. . . .[1]

A deed filed in Trinidad, county seat of Las Animas County reads as follows:

> This Deed, made this twenty fourth day of July in the year of our Lord, one thousand eight hundred and eighty three between Edward West, Town of Trinidad of the County of Las Animas and State of Colorado, of the first part, and John A. Thatcher, Mahlon D. Thatcher and Frank G. Bloom known as the firm of Thatcher Bros. & Co. of the County of Las Animas and State of Colorado, of the second part.
>
> WITNESSETH, That the said party of the first part, for and in consideration of the sum of Two Hundred Dollars, to the said party of first part, in hand paid by the said party of the second part, the receipt whereof is hereby confessed and acknowledged, has granted, bargained, sold and conveyed, ... unto the said parties of the second part, their heirs and assigns forever, all the following parcels of land, situated, lying and being in the County of Las Animas and State of Colorado, to wit.[2]

The tracts conveyed contained 480 acres and probably had access to surface water.

The dawning of the 1880s saw the Thatcher brothers and the First National Bank of Pueblo pushing back their horizons. The bank was extending credit to an ever-growing number of stock growers and other industries in Colorado and beyond. The brothers spent most of their time overseeing their expanding banking business, managing their growing mercantile and mining enterprises, and keeping in touch with their growing investments in cattle with the H. W. Cresswell and Company branding the Bar CC, the Anderson Cattle Company branding the Diamond A, and their own cattle operation branding the Circle Diamond.

Frank Bloom rode herd on the Circle Diamond livestock in Colorado, where their range operation had spread from the Hole-In-The-Rock Ranch. Circle Diamond cattle and horses watered on the Thatchers' scattered parcels of deeded land and grazed the adjoining expanses of public domain. From the headquarters near the new Atchison and Topeka Railroad siding named Thatcher and from line camps set at other primary waterings, mounted cowboys kept a watchful eye on the thousands of Circle Diamond cattle and horses that ranged west along the Apishapa and Timpas creeks in the Bent and Lockwood canyons country.

In May 1882, grass-fat cattle brought $9.35 a hundred liveweight on the Chicago market. Three- and-four-year-old steers

Broodmares on the Circle Diamond Ranch near Thatcher, Colorado, circa 1885.

Chuck wagon, Box Ranch, circa 1927. Courtesy Colorado Historical Society, Trinidad Collection.

Original Santa Fe Depot at Thatcher, Colorado. A. R. Mitchell Collection, Colorado State Historical Society, Trinidad.

ready for fattening on the cool-season grasses of the northern range of Wyoming and Montana could bring as much as fifty dollars a head on their home range in Texas and sixty dollars when delivered to Wyoming. The cost to investors looking to cash in on these profit-producing prices was a twenty-dollar investment in a two-year-old Texas steer.[3]

John and Mahlon's partnership with Perry Baxter and Henry Cresswell in the Cresswell Land and Cattle Company, which had succeeded the H. W. Cresswell and Company after relocating the operation to Texas in 1877, was prospering under the watchful eyes of Cresswell. The range count of Bar CC cattle had grown in five years, by purchase and natural increase, to more than thirty-one thousand head in 1882.

Canny investors in Edinburgh, Scotland, hearing about the large profits that were being made from cattle ranching on the Great Plains, sent their manager Murdo Mackenzie to Trinidad, Colorado in 1880 to establish the headquarters office for the Prairie Cattle Company, Ltd. Records show that the syndicate was backed by seven hundred thousand pounds sterling with which to buy land and cattle.[4] Sometime in 1882, Mackenzie, recognizing the potential for profit in a cattle operation in the Texas Panhandle, approached the partners of the Cresswell Land and Cattle Company with an offer to purchase the Bar CC livestock and range holdings. They were unable to reach a meeting of the minds and the deal did not materialize.

By 1885, thousands of Bar CC cattle and over five hundred head of horses were scattered over parts of four counties in the northeast corner of the Panhandle. The company had patented acreages and claimed range rights from the drainage of the Canadian River to the Indian Territory in Oklahoma, more than 1.25 million acres; over 1,953 square miles of pristine, prime grazing land.[5] In order to be more centrally located, Cresswell moved his headquarters from Home Creek to near Wolf Creek in Ochiltree County. He purchased from Al Barton the old Ed Jones and Joe Plummer buffalo hunters' outpost.[6]

Pioneers in the business of raising cattle, who had experienced the cyclical nature of supply and demand, weather, and markets knew that they could take advantage of inexperienced investors by selling out their herds, their few acres of patented land with stock water rights, and their nebulous range rights, all at their price, on a seller's market. Or they might elect to exchange their cattle and range rights, in which they had little or no initial investment, for stock options in a company with a bulging purse eager for further expansion.

In May 1885, the following notice appeared under the legal notice section of the *Star Journal (Pueblo, CO) Chieftain*:

KNOW ALL MEN BY THESE PRESENTS, That we, John A. Thatcher, Mahlon D. Thatcher, Frank G. Bloom, Wilbur F. Bloom and John Burns, residents of the State of Colorado, have associated ourselves together as a company under the name and style of "The Bloom Cattle Company," for the purpose of becoming a body corporate under and by virtue of the laws of the State of Colorado, and in accordance with the provisions of the laws of said State, we do hereby make, execute and acknowledge in triplicate this certificate in writing of our intentions so to become a body corporate, under and by virtue of said law.

First. The corporate name and style of our said company shall be "The Bloom Cattle Company."

Second. The object for which our said company is formed and incorporated is for the purpose of raising and dealing in cattle and horses and carrying on a general stock business in the Counties of Bent and Las Animas, in the State of Colorado. And it is the purpose of the said company to carry on and conduct part of its business beyond the limits of the said State of Colorado, to-wit: in the Territory of New Mexico.

Third. The capital stock of our said company is five hundred thousand (500,000) dollars, to be divided into five thousand (5,000) shares of one hundred (100) dollars for each share, and said stock shall be nonassessable.

Fourth. Our said company is to exist for the term of twenty (20) years.

Fifth. The affairs and management of our said company to be under the control of five (5) directors, and John A. Thatcher, Mahlon D. Thatcher, Frank G. Bloom and John Burns are hereby selected to act as said directors, and to manage the affairs and concerns of said company for the first year and until their successors are elected.

Sixth. The operations of our said company will be carried on in the Counties of Bent and Las Animas, in the State of Colorado, and the principal place and business office of said company shall be located in the City of Pueblo, County of Pueblo and State of Colorado aforesaid.

Seventh. The directors shall have power to make such prudential by-laws as they may deem proper for the management of the affairs of this company, according to the stature in such case made and provided.

IN TESTIMONY WHEREOF, we have hereunto set our hands
and seals on this 26th day of May, A.D. 1885.
John A. Thatcher   Seal
Mahlon D. Thatcher Seal
Frank G. Bloom   Seal
Wilbur F. Bloom   Seal
John Burns   Seal

John Thatcher, in the vernacular of the cowboy's world, rode
point for The Bloom Cattle Company; or in the corporate board
room of the twentieth century, he was the CEO. Mahlon was secretary and treasurer. Frank was general manager. The three-
member team and their associates were laying the groundwork for
a ranching conglomerate that would become one of the largest
range cattle, sheep, and horse operations in the history of our livestock industry in the United States. Their participation in the development of agriculture in the Great Plains had a positive
influence that is still evident in the closing years of the twentieth
century.

CHANGING TIMES

The decade of the 1880s was another time of transition on the Great Plains. Now the windmill supplied water for the granger's family and stock, and his plow and harrow turned the sod under and prepared a seedbed. Miles of barbed wire stretched between wooden posts cut the open range into claimed parcels. The existing laws for the sale and disposition of public domain permitted entry by individuals, associations, and corporations. Under the several settlement and occupancy laws, a person could legally acquire 1,120 acres of public domain by residence, improvement, and cultivation. Agricultural lands could be purchased for a minimum price of $1.25 per acre. These changes began to affect the cattle companies, whose operations were dependent on range rights to free pasturage on public land.

In 1884, the Thatchers could see the tidal wave of change moving westward across the prairies of Kansas and Nebraska toward the eastern plains of Colorado. With an awareness of the new day at hand they began expanding the base of their far-flung livestock operations. To the grazing operation headquartered at the Hole-In-The-Rock Ranch in Las Animas County was added some three thousand acres in scattered parcels of patented and leased land. And they continued to increase their land base between the Arkansas and the Huerfano Rivers in Pueblo County. With the backing of John and Mahlon, Cresswell was acquiring patented land in Ochiltree and Roberts Counties in Texas, and Anderson was gathering title in the name of the Anderson Cattle Company to all the land he could lay his hands on for thirty miles along the Rio Hondo in Lincoln County.

A letter postmarked December 27, 1883, in Austin, Texas, on Hotel Brunswick stationery reads as follows:

John A. Thatcher

Dear Sir

Yours of the 23rd I rec'ed yesterday, just as I was leaving for Dennison, I take this the first opportunity of answering it. In regard to buying at the prices mentioned you need not worry about that. I am down here and will see what can be done and if anything it will be at such figures as will leave a reasonable margin for profit, the glory part of this business I don't bank on. I took some land Gunter had bought for us at $1.25 but told him we would only pay $1.00 in future with the exception of River land two Sections of which I agreed to pay $1.75 for & some other land I bid 80¢ for. He, Gunter, thinks the land we contracted for will all or nearly all be patented in two weeks time. Since I have been down at Sherman I find there have been sales made at $12.00 & 16.00 for 1s & 2s at San Antonio. I will not pay more than this and don't have any idea I will pay that much. These fellows here are the damndest frauds in existence. If a stranger comes down here they all plot against him and while very sociable work against him all the time, our only chance will be catching some fellows hard up which I think may be done as money is tight and but few buyers in the country. Will write again soon.

Yours truly,
H. W. Cresswell[1]

A letter addressed to Frank Bloom in Trinidad states:

Mr. Thatcher writes me that he will be down again the 20th or 25th I hope you will come down with him. Two of the Mexicans above me on the River are hard up for money and have sent me word they will sell out to me. They own farms adjoining the upper end of my range. I have not seen them yet, but have waited a long time for this thing to turn up and just enough time for me to either buy them out, if I can do so reasonably, or else advance them some money, on interest and take a mortgage on their places. This I would rather do as they will never be able to pay the money back.

Yours Resp.
W. E. Anderson[2]

The Thatchers' original brand was the Circle Diamond. Through purchases or foreclosures whereby the Thatchers might have acquired the rights to brands along with the livestock, the company registered and claimed livestock under various other brands.[3]

In 1885–86, grazing cattle on public domain was still the method of operation for the majority of large investors. To the already crowded late-summer ranges of Kansas, Colorado, and the Texas Panhandle, and to the glutted market of 1885, were added some two hundred thousand head of cattle that were forced off the Cheyenne and Arapaho reservation by a proclamation from President Cleveland.

The big ranching companies that were monopolizing the western ranges were often under the administration of absentee owners who had no experience with range life and animal husbandry. Their primary concern was the return on their investments. Though their ranching ventures were dependent on the bounties of nature, the investors were, for the most part, uninformed and oblivious to the fragile ecology that exists between growing plants, weather conditions, and the weight-of-harvesting on the Great Plains.

The appropriation of public lands in the early days of ranching was based on the principle that might makes right. The investors' might was created by the power of money, and their right was enforced by the Winchester rifle and barbed wire. The Coad Ranch sold in western Nebraska for $912,853. The legal inventory described in the sale contract included the right of occupancy to most of one county plus "a main pasture of 143,000 acres," which the owners claimed to "have no title except a possessory title or right thereto." According to the records, they signed over the title to a total of 527 patented acres.[4] In Colorado, the SS Ranch sold to the Arkansas Valley Land and Cattle Company of London, England. The actual deeded land transferred in the sale contract was only 14,147 acres. The balance of the property was "about seven hundred thousand acres of unclaimed land belonging to the government of the United States." The price paid for patented and "unclaimed" land under fence, some 8,660 head of cattle and the equipment was $707,680.[5]

The cattle barons, who claimed range rights to vast spreads of free grass on public lands, were destined to go the way of the great herds of buffalo and the nomadic Plains Indians. The practice of large operators fencing in their range rights was being contested by other stockmen and by families from the east who listened to government propaganda and thought they could make a living farming a homestead on the High Plains west of the hundredth meridian. The response from the secretary of the interior was that the government did not sanction the privileged use of public lands by individuals or corporate entities, and therefore, fencing public lands for exclusive use by an individual or corporation was not legal.

A letter address to Wilbur Bloom, Frank's brother and foreman of the Circle Diamond Ranch:

August 11, 1885

Wilbur Bloom, Esq.
Thatcher, Colorado

Dear Sir:

Herewith I enclose a printed copy of a proclamation issued by the President yesterday on the question of taking down fence enclosing Public Domain. In consequence of this order I think you had best go to work at once and take down your fences.... Please advise me what you have done and how soon you will be through taking down that fence.

truly yours.
Jno A. Thatcher

Ranchers retaliated with threats and violence against the cutting of their perimeter fences and the intrusion of cattle not wearing their brands. The owner of cattle that had been herded past cut wires in a fence onto another man's range might find when he went to gather them that his bulls had suffered the indignity and discredit of castration and his tight-bagged cows were bawling for their missing calves.

A war broke out in Texas between the family ranchers and the corporate ranchers. Gunmen were put on the payroll to prowl outside fences and track down those who used wire cutters to destroy miles of fence that was erected to enclose range rights claimed by large operators. The Wyoming Stock Growers' Association, with backing of local and state politicians, most of whom grazed their livestock on the open range, paid territorial funds to range detectives to arrest, and in some cases do bodily harm to anyone who broke the rules and regulations of the Association, which included rustling Association members' livestock and cutting their fences. A vigilante band, dubbed the Stranglers and said to have been financed by a clandestine group of prominent northern cattlemen, found that punishing a man proven guilty of stealing grass in the accepted manner of dealing with horse thieves, by hanging, gave pause for others who might be looking to move against another man's range rights.[6] According to his biographer, Charles Goodnight took the stance that "free grass on public lands and the six-shooter went hand in glove."[7]

The changing of the guard prompted other more judicious ranchers to yield to the forces of transition by taking steps to acquire title to land. By fair means or foul they sought out acreages that would either help establish a more permanent base for their livestock operation by forming a legal barrier to public lands they were using, or provide stock water on patented land for cattle or sheep that were utilizing public domain, thereby laying claim to range rights for their meandering livestock. A company's hired men or some disinterested party might be encouraged to file on homesteads within the company's pasture and then relinquish the entitlement to the company for a small consideration. Or land could be acquired by the purchase of a harassed, discouraged homesteader's filing. Here again, money was the leverage for attainment.

The Circle Diamond and the Bar CC herds continued to suffer heavy losses from the winter storms that swept out of Canada to scourge the eastern prairie of Colorado and the Texas High Plains. Contrasting these losses with reports from W. E. Anderson, who had been wintering the Diamond A cattle along the Hondo River in New Mexico with no losses due to inclement weather, had prompted John Thatcher to send Frank Bloom south in search of more favorable breeding range for Circle Diamond Cattle.

When Frank Bloom returned from an inspection trip of Lincoln County, New Mexico, he wrote to John Thatcher.

Trinidad, Colo.

Sept. 17, 1884

Jno A. Thatcher, Esq.
Martinsburg, Pennsylvania

Dear Sir:

I sold our beef on Tuesday to Wayman of Emporia for 40$ delivered Nov 1st. I got to studying about the lower ranche and hearing that Billy Wilson of Durango and another party from there was here yesterday on their way south to look up a ranche.

I will leave for Anderson's tomorrow and go to San Antonio and Ft. Stanton and buckboard on to the Hondo. To go from here by wagon and get back by Nov. 1st I haven't time enough. Will take my robes, blankets, rifle etc with me. By going this way it will give us 20 days time from the time I get there until Anderson has to ship his beef and that will be long enough to look over a great deal of that country. My plan is this in the six weeks time I now have. Can give the Sacramento Mts a thorough looking over and if they do not suit I will take my wagon and team with a good party and strike out but I got afraid to risk this any longer. I got new Map of 1883 which is better than the one we saw as it shows the Guadalupe mountains on the south of the Sacramento with a big valley between. I will write you when I can to Pueblo. Hoping to be lucky on this trip and get in there ahead of any one. I am,

Yours Truly,
Frank G. Bloom

A handwritten reply dated July 6, 1885 from John to Frank:

I am quite sorry to hear that the Thurber and DeLancy outfit are getting hold of those ranches above us on the Hondo. It leaves us but one resort and that is to buy such of those places as will serve us best and we must not lose too much time in doing so. We must only buy such places as will be of the most use to us in the way of water and outlet range.

On July 12, Frank received a follow-up letter from John:

[I]t is important for us to get a footing on the upper Hondo, but it is also important to get it without too much of an outlay of money. I will leave this now to your own judgment as you are on the ground and will realize the situation better than me and can appreciate what to do. Anderson is deposed to get wild at times about ranches and likely not to

use sufficient judgment in making purchases. So I will leave the matter
to your judgment and know you will advise Anderson for what you
think best.

On July 29, John addressed a letter to Frank at Roswell:

Yours at hand. Glad to hear from you. From your letter and from one
received from Anderson several days since I should suppose that the
Lee DeLancy, Thurber & Co. are getting wild about ranches on the
Hondo. They must be exceedingly foolish to show their hands in such
an open manner. There is one thing which I wish to call to the attention of yourself & Anderson especially which is this. I think about all
the ranches owned or claimed by Mexicans on the Hondo have not as
yet had their titles perfected by the Government and had no patent issued for them. Now in buying this class of places, if you buy any in that
fix, I would not pay but a very small sum of money on them if any before the title is perfected and that may require quite a long time. This
idea of buying their claim and the chance of getting a perfect title in
time I consider unwise and likely to create serious trouble which we do
not want, beside losing the money paid out. Mr. Hopkins tells me that
the Gen land office agent in Rocky Ford is quietly looking up every
quarter section of land in that country owned by cow men, asserting
how the title was obtained & reporting it to Washington. If this is carried out thoroughly it will destroy the title to many of the ranches in
this part of the country.

At the time the Bloom Cattle Company was looking to branch
out and acquire property along the Hondo Valley, Lincoln
County was the largest county under the Stars and Stripes. Covering almost the entire southeast quarter of New Mexico, it was
about equal in land area to the entirety of Vermont, New Hamp-
shire, and Massachusetts.

Since the 1860s, John, Mahlon, and Frank had been trailblazers
on the frontier along the Front Range of the Rocky Mountains.
Their personal lives and business enterprises had been influential
in shaping the society and character of their adopted home. In
1880, Colorado had a registered population of 194,320 and was the
most populous state in the Mountain West.

The directors of the Bloom Cattle Company were about to em-
bark again into pioneering on another frontier. In the 1880s, the
territories of New Mexico and Arizona had all the earmarks of an
uncurried wilderness. The vast region was violent and unruly under the guns of men riding roughshod over the populace. The

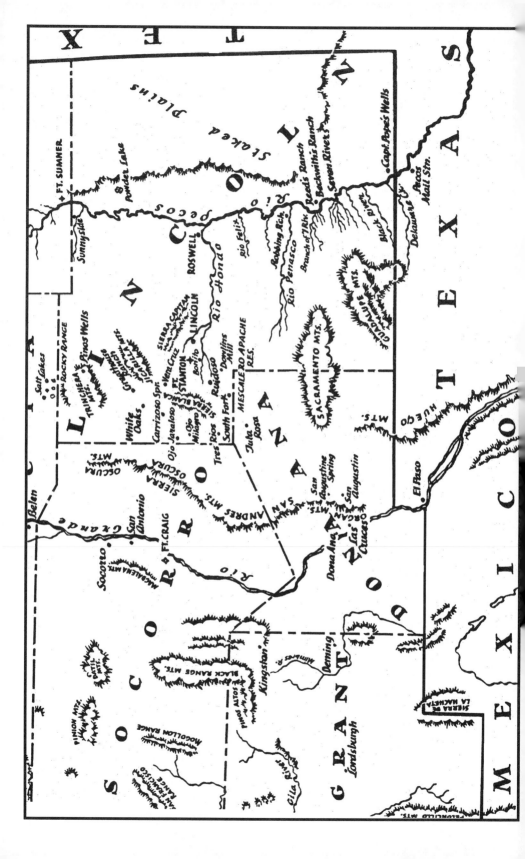

Texas Rangers had been crowding the shirttails of wanted men listed in the *Texas Crime Book*. Many of those hunted men, feeling the heat and looking for refuge and a climate more agreeable with their lawless ways, skipped across the border into New Mexico and Arizona.

Rustlers rode the ranges bordering the Pecos and Canadian Rivers with a running iron, used for branding the unbranded and altering brands, tucked under their stirrup leathers. In the eyes of tough old cowmen, the packing of a running iron was in itself incriminating evidence, the mark of a rustler. And it may have been that they had a first-hand basis for their convictions, as many had started their own herds by swinging a wide loop and packing a running iron.

In the spring of 1885, Frank Bloom was supervising the roundup, sorting, and loading of some twenty-five hundred Circle Diamond cows, heifers, and bulls into Atchison, Topeka and Santa Fe (A.T.&S.F.) Railroad stock cars spotted beside the corral on the siding at Thatcher, Colorado. The waybill showed the destination for the cattle was the wildest of the wild west towns, Las Vegas, New Mexico.

At the same time, W. E. Anderson's cowboys were riding the vast, open range that flanked the Hondo Valley rounding up the two- and three-year-old Diamond A steers. When the gather was completed, the cowboys rolled their soogans, a supply of smoking material, their BVDs, socks, an extra shirt and trousers, some whang leather for saddle and tack repair, an extra girth, and their shaving gear into tarpaulins, or ground sheets, that protected their bedding. They cinched up their bedrolls with spare lariats and loaded them onto the hooligan wagon. The cook loaded provisions and his cooking gear, filled the water barrel, climbed up on the plank seat of the chuck wagon and cracked his whip over the backs of two spans of mules. With a point man in the lead, swing men on both sides and the drag men behind, the Diamond A herd took the trail to Las Vegas, where the steers were to be loaded on the A.T.&S.F. and railed to the livestock market in Kansas City.

It had been arranged by correspondence in May that the Circle Diamond cowboys would hold the breeding cattle on the meadows around Las Vegas until after the Diamond A steers had

Hooligan Wagon, Circle Diamond Ranch, Colorado. Courtesy Colorado Historical Society.

trailed in and been shipped. Then, under the direction of Frank Bloom, the Diamond A crew would string the Circle Diamond herd out, backtracking alongside the Pecos River to beyond Fort Sumner. Leaving the Pecos Valley they would trail the cattle west along the north side of the Capitan Mountains, watering the herd in spring-fed creeks that fanned out of the canyons onto the prairie. At the little Mexican settlement of Las Tablas they turned the herd south following the Bluewater and Escondido Creeks through the rugged Capitans to the headquarters of the newly established Bloom Cattle Company Ranch in the Hondo Valley.[8]

WHITE CHRISTMAS
ON THE PLAINS

The winter of 1884–85 had been rough on livestock in the Texas Panhandle. As Cresswell reported to John Thatcher in a letter dated March 3, 1885: "There is a fearful loss of cattle in Texas. They think 40 per cent. They are dying by the thousands now although the grass has started in two weeks however they will be all right."

The general practice was for ranchers to stack enough prairie hay at the headquarters and outlying camps as a feed supplement for the saddle horses and milk stock during severe winter weather. It was expected that beef cattle and the horses that were turned out for the winter would make it until spring on what they could rustle.

When winter winds howl and the mercury in the thermometer hovers around the zero mark, animals instinctively seek the protection of sheltered areas to wait out the storm. Livestock caught on the open prairie are at the mercy of the elements. Horses will stand with their butts against the wind; bison will turn their woolly heads into the teeth of the gale and eventually walk beyond it; but cattle will drift along with the storm. When snow blankets the prairie, the horse, with its cup-shaped hooves, is able to paw through the ground cover to get to the grass, but cloven-hoofed animals like cattle and sheep do not have this capability.

In the days before ranchers on the Great Plains faced up to the reality that their pastured animals needed supplemental feed during the winter, the howling winds of a blizzard could be the death knell for livestock caught on the range. It was a delicate balance, matching the animal's condition and vitality against the elements and time. When the grass was buried under deep snows capped with ice and the oblique rays of the sun in the southern quadrant never warmed the daytime temperatures above freezing, then only the coming of a timely chinook with its warming winds offered cattle a reprieve from slow starvation.

The day when cattlemen and speculating investors hired cowboys to trail in great herds of cattle to multiply and fatten on the grasses of the southern Great Plains was coming to a close. A time was at hand below the North Platte River when the demand for cattle no longer exceeded the supply and when the 20½ percent dividend paid by the Prairie Cattle Company in 1883 to stockholders would be only a memory.[1]

The administration and the management of ranches would change, out of necessity. Administrators would find themselves saddled with more than just a concern over investment strategies and returns. Ranch and herd management became a hands-on responsibility. Owners and stockholders would learn that the accounting of herd performance could not be left to employees whose jobs depended on favorable reports. Differences between cattle on the hoof and inventories given to investors could no longer be glossed over, nor losses due to poor management be shrugged off on the premise that time and reproduction would make the inflated count a reality. Administrators of ranching companies learned to understand the validity of the old truism: "It's the master's eye that fattens his cattle." They would spend less time behind a rolltop desk and more time on the range, callusing their rumps on the seat of a saddle or a jouncing buggy and dirtying their boots by walking in fresh cow patties.

After negotiations for the sale of the Cresswell Land and Cattle Company to the Prairie Cattle Company failed in 1882, the company continued to expand their herds and land holdings. Cresswell purchased three thousand head of cattle branded P O and the adjunctive range rights below the South Fork of the Canadian in Hemphill County from Milton Pollard. After his friend and neighbor, Joseph Morgan, died of smallpox in 1883, Cresswell paid the widow ninety thousand dollars for three thousand head of Triangle (△) branded cattle and her range rights in Lipscomb County. The Bar CC partnered with E. T. "Nick" Eaton on the U Bar U range along the North Fork of the Red River in Gray and Wheeler Counties. They branded the Forked Lightning (⤛) on these cattle.[2]

The perennial bachelor Henry Whiteside "Hank" Cresswell was, according to historical records, a remarkable individual, an example of the Thatcher brothers' ability to judge character and attract above-average associates. E. H. Brainard went to work for Cresswell on the Bar CC in the fall of 1883. In a letter dated July 19, 1926 to J. Evetts Haley, Brainrd stated: "Cresswell was a fine man. He was exceptional in many ways, and liked by everybody." The Panhandle Stock Association was organized in the spring of 1880. Charlie Goodnight and Hank Cresswell were the first elected president and vice president, respectively.

In the days before the necessity of written contracts, Cresswell's word was his bond. Judge O. H. Nelson contracted to buy some two thousand head of Bar CC cattle. Cresswell was ill and not able to be on hand when the cattle were delivered to the judge in Dodge City. He sent word with the trail boss for Nelson to accept the cattle on the condition of their agreement; and they would settle up later. Nelson held the sixty thousand dollars for more than a month without hearing from Cresswell. They finally chanced to meet and settled the account. There was no mention of interest. Nelson was reported to say: "They didn't make 'em better than Hank Cresswell."[3]

The Scottish stockholders of the Prairie Cattle Company still wanted to acquire the Bar CC. They interested some English investors in the project and approached the partners of the Cresswell Land and Cattle Company again. In January 1885, the terms of a trade were finalized. The Prairie Cattle Company bought controlling interest in the Bar CC for 1.5 million dollars. A new company, composed of the stockholders of Prairie Cattle Company together with the owners of the Cresswell Land and Cattle Company united to form the Cresswell Ranche [sic] and Cattle Company. The partnership of John, Mahlon, Perry, and Cresswell retained a stock investment valued at two hundred thousand dollars in the new company. Cresswell agreed to remain at the ranch to see to the interests of his partners until the terms of the trade were fulfilled and to assist with the transition. In June 1885, Mahlon went to Scotland to handle the exchange of title and money. Included in the negotiations were fifty thousand head of cattle, thirty thousand acres of patented land and seventy thousand acres for which patent had been applied.[4]

The death losses during the severe winter of 1884–85 started the transfer of ownership off with a problem. During the roundups, Cresswell began to find the cattle count on the ranch coming up short of the fifty thousand cattle he had guaranteed.

In a letter from Austin, Texas, postmarked March 3, 1885, Cresswell advised John Thatcher:

Dear Sir,

I have had a talk with Barton and have concluded to take about 3 or 4 thousand head of western cattle at not over nine dollars per head started on the road. This with decent luck would put them on the Panhandle for less than $10.00. These are the top cattle in the State. Barton had bought most of the horses at low figures and I thought as we had a remnant to bring we might as well have a herd. I told Barton if he could get these cattle at this price and of course tops if possible I would take them but not one nickel more. I today telegraphed you to place $10,000 to credit of Clayton Barton at the First National Bank of Austin, Texas the same bank as last year. Barton says he saw Goodnight who told him I had sold but said if you see Hank tell him from me if he has sold and wants to buy tell him I will take care of 15,000 head for him, of course I would have to see Goodnight about this to find out really what it means. We might be able to take some steers to the Arkansas River and hold them a couple of years if we saw fit. We can talk this over when I come up.

Yours truly,
H. W. Cresswell

Cresswell purchased some twenty-two thousand head of cattle in Burnet and Lampasas Counties. The cattle were thrown on the road, headed for the Bar CC range in the Panhandle, a trail drive of some four hundred miles.

Cresswell, aware that the intrusion of an additional twenty-two thousand head of cattle would put pressure on the Bar CC's use of public lands, rode ahead to confer and make a deal with Goodnight for pasturage. An agreement was made and Cresswell cut some eleven thousand head out of the herd and drove them to Goodnight's JA ranch. The combined crews of the Bar CC and the JA put the cattle through a chute and burnt a Lightning brand on the left ribs. Those were cattle belonging to Thatcher-Cresswell-Baxter partnership. They were turned loose to run in Palo Duro Canyon. The balance of the twenty-two thousand head were driven on to the Bar CC headquarters at the head of Wolf Creek where they were branded with a Chain H for the Cresswell Ranche and Cattle Company.[5]

The cattle summered well on Wolf Creek and went into the winter in good flesh. However, during the fall work, cowboys riding the range observed that the Chain H cattle from south Texas were still wearing their summer coats while the native-raised cattle and horses were growing an unusually heavy coat, a sign from nature portending a severe winter. The Chain H cattle went into the winter of 1885–86 foraging on the open plains and in the brushy ravines between Wolf Creek and the North Fork of the Canadian River.

While the winter of 1884–85 had been bad, it was but an omen of the deadly fury possible from winter storms on the High Plains where, according to men who fought the elements, "there's nothing between the Panhandle and the North Pole but a 4-strand barbed wire fence." True to the forecast of the heavy hair coats, the winter of 1885–86 was, in the words of the cowboys who suffered through it, "a ring-tailed doozy."

The first blizzard hit on January 7. Thousands of Bar CC cattle were driven along by the force of the northerly winds to hump up, bawl and die along the north side of a two-hundred-mile, four-strand, barbed wire drift fence that Cresswell and neighbors had stretched along the Canadian River breaks at the south end of the Bar CC range. Of the approximately eleven thousand head of Chain H cattle, Cresswell tallied some eight hundred still on their feet after the spring thaw, something like a ninety per cent loss. Fortunately for the Thatcher-Cresswell-Baxter partnership the cattle wearing their Lightning brand were sheltered in the depths of Palo Duro Canyon and their losses were light. Adding to the disastrous losses to Cresswell Ranche and Cattle Company's cattle during that winter were losses to predators during the spring and early summer calving season. In a letter to Pueblo in the late spring of 1886, Henry wrote that on parts of the Bar CC, "wolves have destroyed more than half of the calves and are in a fair way to finish them all."[6] Cows that survived and calved in the early spring were in such poor condition that many of their calves either died of starvation or were dogied.

> The dogie's a calf without parents or friends.
> He's fat in the middle and poor at both ends!
>
> *Rawhide Rhymes,* by S. Omar Barker,
> whose brand was the Lazy S O B (⌐ᴜOB).

Compounding the cattlemen's problems created by the winter death losses, estimated at near eighty percent, and by the steady losses to predators and rustlers, were the drought and a plague of grasshoppers that put a stranglehold on the Great Plains during the summer of 1886. A scant cover of prairie grasses lay brown and dormant, as in the dead of winter, and springs and creeks were drying up. Grass fires during the late summer and fall set by lightning, and possibly by Indians, left a vast area of the Bar CC charred and dreary.

Cattle prices declined steadily as nervous investors instructed their ranch managers to round up the cattle and rush them to market. Slaughter cattle on the Chicago market brought from $2.80 to $3.00 per hundred pounds, live-weight. This was ten to fifteen dollars less than a year earlier.[7] An editorial in the August 26, 1886, issue of the *Rocky Mountain Husbandman* warned: "Beef is low, very low, and prices are tending downward, while the market continues to grow weaker every day. But for all that, it would be better to sell at a low figure, than to endanger the whole herd by having the range overstocked."[8]

The market for stocker cattle, from the Front Range of the Rocky Mountains to the western banks of the Missouri River, was as depressing as the lackluster prairie. Cresswell sent a herd of some three thousand three- and four-year-old steers up the Jones-Plummer Trail to market in Dodge City. When Cresswell learned that the markets were glutted with cattle and prices were plummeting, he sent a rider at a gallop after the herd with word to turn the cattle around and drive them back to the ranch.

Rains across the Panhandle in 1886 had been disappointing. Old-timers were shaking their heads and muttering about the outlook for winter feed. To the already crowded public lands in the northern Panhandle came cattlemen looking for grass for herds numbering in the thousands, which were part of the two hundred thousand head that had been removed from the Cheyenne-Arapahoe reservation by proclamation of President Cleveland. Due to the lack of moisture during the growing season in 1886, the stunted grasses went dormant before making seed and cows nursing calves were in poor condition. Many of the surviving calves went into the fall as pot-bellied, dead-haired dogies.

Cattlemen from Montana to Texas, recalling the past two winters, were apprehensive over the old notion that calamities come

Sorghum bundles for supplemental cattle feed on the Great Plains.

in threes. On most ranges the scant summer rains had not produced enough pasturage on which to sustain herds until hoped-for new growth in the spring of 1887. But to liquidate at current prices assured a financial disaster. Some of the northern herds were driven into Canada, where it was possible to lease tracts of land from the provincial government. Desperate cattlemen looked across their overgrazed pastures at vast tracts of grassland in western Oklahoma, which had not been grazed heavily since the days of the migrating buffalo in the late 1860s. Some cattlemen with political clout received permission from the Department of the Interior to negotiate leases with the tribes for grazing rights on the reservations.

Living in the building that Jones and Plummer had built some years ago as a dwelling and storeroom for buffalo hides, Cresswell was at the Bar CC headquarters during the time it was taking to transfer ownership of the Cresswell Land and Cattle Company to the Cresswell Ranche and Cattle Company. Riding over the droughty range the Cresswell Land and Cattle Company was leasing from Charlie Goodnight, Cresswell knew that he had to find a new home for the eleven thousand Lightning-branded cattle in Palo Duro Canyon. Taking advantage of the fire-sale market for all classes of cattle, and gambling on a better market in 1887, the Thatchers, Cresswell, and Baxter bought an additional eleven thousand head of steers from Goodnight. Late in the summer of 1886, he managed to lease grazing rights and move the twenty-two thousand head of cattle into Indian Territory known as the Cherokee Strip between the North Fork and the South Fork of the Canadian River in the Oklahoma Panhandle.[9] There was a good stand of old grass and good protection against winter storms in the drainage breaks along and between the valleys of the two rivers.

The winter of 1886–87 separated the dyed-in-the-wool cattlemen from the board-room speculators. That fall, Indian summer ended in a mass of cold air and a heavy snowfall that blanketed most of the Great Plains. A series of storms kept the grass covered and sent shivering, gaunt cattle to huddle against cut-banks and in the timber belts seeking protection and something to eat.

The haystacks of even the more astute stockmen were depleted before Christmas. A chinook in early January raised hopes for a reprieve, but they were dashed by a prolonged blizzard that started on January 28. It roared in from the Arctic across Canada and on over the Great Plains. It howled for three days. The mercury in the thermometer hovered at the bottom of the stem for weeks. The snow became a hard, ice-capped mass that lacerated and froze cattle's feet. An animal's body heat would thaw the snow on their backs during the few hours of sunlight, leaving icicles hanging from their sides like long glass baubles. When the exhausted critters sought rest or protection from the chilling winds and lay down, their wet coats would freeze to the ground.

Weakened and unable to break free, they died where they lay. Cattle and horses ate willows to the ground and stripped the bark off the cottonwood and aspen trees. They pulled what branches they could reach off the trees. Horses ate each others' tails and manes. The February 2, 1887 issue of the *River Press* out of Fort Benton, Montana, carried the news: "The temperature at 2 o'clock today was 42 degrees below zero, with a gale blowing from the north." This was before windchill factors were computed, but not before the killing effect of the combination of bitter cold and biting wind was a cruel fact of life. Not even the old pioneers could recall such a punishing winter on man and beast alike. It was the third punishing winter in as many years.

Riding the ranges after the spring thaw of 1887, cattlemen across the Great Plains found the death losses were staggering, especially in open, level country which lacked browse and natural protection. Carcasses were piled up by the thousands against cutbanks and in ravines where wind-driven cattle had sought protection from the killing blizzards. Fences that held against the pressure were lined on the north sides, mile after mile after mile, with the frozen bodies of cattle that had tried to drift along with the Arctic winds.

Charles M. Russell, the cowboy artist, during his early days as a cowboy, was in a line camp in the Judith Basin of Montana during that disastrous winter. His employer sent word inquiring about the condition of the cattle. Russell's reply was a sketch in water color of a pack of hungry wolves eyeing a solitary, humped-up, starving cow standing in the snow. His caption was: "Waiting for the chinook."[10] Under the title *The Last of the 5,000*, it became one of his classics.

The impact of changes on the cattle business on the Great Plains during the decade of the 1880s came to a devastating climax with those three killing winters, back-to-back, and the collapsing cattle market. Many of the large outfits found that their companies were built not on solid rock but on sinking sand. Investors from the money capitals in the eastern states and abroad who had been eager to cash in on the reported sure-fire, big profits to be made on cattle in the West had not bothered to learn the intricacies and volatility of the ranching business before they took the plunge.

Capitalists, prudent men of industry, who had made their money by shrewd investments with a firm hand on the tiller, jumped headlong at the chance to sink their money into cattle ventures on public lands. Most speculators had not the vaguest idea of the rudiments of ranching. A buyer would purchase, usually through a commission man and sight unseen, thousands of head of cattle based solely on a seller's statement as to the number and classification of cattle wearing the seller's brand and a rough description of the range rights that went along with the cattle. Most absentee owners, sitting in their offices reading reports from hired men, did not know much more about their cattle operations than did the stranger offering to buy. Accepting the seller's tally as correct and tendering payment on a per-head basis for the different classifications of cattle, the buyer would turn his nomadic properties over to a crew of dollar-a-day cowboys. After a buggy ride to the ranch headquarters and a quick tour during which grazing cattle raised their heads to look without much interest at their new owner, and then several scotch-and-sodas with other cattlemen in the Cheyenne Club in Wyoming or the Shirley Savoy Hotel in Denver, the newly minted "cattleman" would board the train and go back east, leaving the reins to his sizable investments in the hands of a manager whose competency had not been confirmed.

Alexander Hamilton Swan came to Wyoming from Pennsylvania in 1873. His Swan Land and Cattle Company herd of Two Bar ($\equiv$) cattle grew to number in the tens of thousands running on public lands between the Chugwater and Sybille Creeks in Wyoming. In 1883, he sold his range rights to a vast spread and his cattle, based on his book count, to a Scottish syndicate for $2,387,675.[11] Pleasure for the buyers from Edinburgh was, as it would be for a Scottish terrier courting a skunk on the prairie of the Great Plains, an educating, short-lived, regrettable affair. The company went into the hands of a receiver in May 1887. It was estimated that the book count on which the Scotchmen based their purchase was short by some thirty-two thousand head.

Records suggest that the Thatchers had suffered financial losses in the disastrous mid-1880s through their involvement with the Cresswell Land and Cattle Company. There is very little evidence at hand to show the Bloom Cattle Company also suffered losses but it is almost a certainty, as records substantiate that cattle losses in southeastern Colorado during those bad winters were severe. According to accounts, Cresswell was not in favor of the sale of the Bar CC to the Prairie Cattle Company. It may have been instigated and pushed to completion by the Thatchers, an economic necessity brought on by heavy winter losses and a poor market.

The inclement weather and poor markets during the last half of the decade of the 1880s, a change of management after Cresswell left the Bar CC Ranch, fluctuating cattle prices for the next few years, and pressures from emigrants coming by rail to take up homesteads in the Texas Panhandle, caused the Cresswell Ranche and Cattle Company to go into a decline from which it never recovered. The company was liquidated around 1900.

THE TURKEY TRACK

The ill winds that had buffeted cattlemen on the Great Plains since the first killing winter of 1884–85 finally blew themselves out in 1889, and a freshening breeze blew across the prairie carrying a promise of better days for the cattle industry. The Cresswell Land and Cattle Company's steers carrying the Lightning brand on the Goodnight lease in Palo Duro Canyon had made it through the fateful year of 1886 with very few losses. The drought broke and the summer of 1887 saw grasses blanketing the prairie that fall. Cattle went into the winter in good shape. Those twenty-two-thousand-and-some-odd head of three- and four-year-old steers that Cresswell put on the Cherokee Strip in the fall of 1886 hit a home run for the Cresswell Land and Cattle Company. They came off the Cherokee Strip in good flesh and hit a rebounding market. Their sales recouped most of the losses the Thatchers, Cresswell, and Baxter had suffered during the past few disastrous years.

From his base in Trinidad, Murdo Mackenzie, manager of Prairie Cattle Company, Ltd., had worked with the Thatchers in Pueblo and Cresswell at the headquarters on Wolf Creek in hammering out the details of the transfer of ownership of the Bar CC Ranch. Mackenzie left the Prairie Cattle Company, in 1886, before the transfer was completed, to head up the huge Matador Land and Cattle Company, Ltd. R. C. Head had taken over as general manager of the Prairie Cattle Company. James McKenzie, an Englishman not related to the Scotsman Murdo Mackenzie, was hired as general manager and took over the management of the newly-organized Cresswell Ranche and Cattle Company. A young cowboy, W. J. Todd, from the Prairie's Cimarron Division was moved to the Bar CC headquarters on Wolf Creek. He was given the foreman's job under James McKenzie and represented the Prairie Cattle Company as he and Cresswell tallied the cattle involved in the change of ownership.[1]

After the cattle were all counted and the transfer completed, Cresswell stayed on at the Bar CC headquarters, acquainting McKenzie and Todd with the ranch and representing the Thatchers', Baxter's, and his own financial interests in the Cresswell Ranche and Cattle Company.[2]

With the completion of the transfer of the Bar CC in 1889, Cresswell and his partners began looking for new territory where Cresswell, the man on the ground, could watch their calves growing into three-and four-year-old, market-age, fat steers. Bankrolled with the proceeds from the sale of the Cresswell Land and Cattle Company, the Thatchers, Cresswell, and Baxter bought into a ranching partnership with A. J. (Tony) Day. They called the company the Cresswell Cattle Company, and branded company cattle with Day's Turkey Track brand ($\wedge$).

Day was born in Texas in 1849. In 1875 he and a brother put together a herd of cattle they had gathered from the wilds along the Gulf of Mexico, branded them with what they called the Turkey Track and drove them to Nebraska. Within a few years they had expanded and were running some fifteen thousand cattle in the Indian Territory of Oklahoma.[3] They dissolved their partnership and Day trailed his Turkey Track cattle into Texas where he became acquainted with Cresswell. His brand, although called by the same name, differed from the Turkey Track brand ($\wedge$) used by the Hansford Land and Cattle Company on their vast ranching operation in the Texas Panhandle and in the Territory of New Mexico.

Mary Terrill, in her article titled "'Uncle' Tony Day and the Turkey Track," published in the June 1943 issue of the *Canadian Cattlemen* "remembered 'Uncle' Tony Day as a stocky man who wore a big white Stetson on the streets of Medicine Hat, Alberta, Canada." She quoted John Clay of the John Clay & Company, livestock commissioners in Chicago, on learning of Tony's death in 1928 as having said that, "Tony Day was one of the three foremost cattlemen of America." Charles Goodnight is reported as having said that Day was a man of Cresswell's caliber. Mary Terrill also stated in her article that "J. L. Peacock, himself an unforgettable personality, referr[ed] to Uncle Tony as 'that grand old cattle monarch' [Peacock wrote] characteristically: 'I have never known a more true, lovable, and noble-hearted man than A. J. Day.... He made friends, only friends, from the Rio Grande to the White

Mud River of Western Canada. . . . I hope that he is resting peacefully in his own kind of country where the grass is long and just enough Indians to keep the settlers back.'"[4]

✻

With most of the open range country of the Texas Panhandle taken up by established cattlemen and grangers packing plows, the newly-formed partnership decided they needed more elbow room for their cattle. Based on their combined experiences pasturing cattle in the Indian Territory, it was decided to lease pasture along the South Canadian in No Man's Land along the Cherokee Strip. Throwing their lots together, the Thatchers, Cresswell, Baxter, and Day stocked their western Oklahoma lease with partnership cattle.

The bull market for cattle that started in the fall of 1889 was brought on by several conditions. Due to heavy marketing during the drought-blistered summer of 1886 and staggering death losses in the winter of 1886–87, as well as a human population growth on the Great Plains, the demand for stocker and slaughter cattle exceeded the supply. With the advent of refrigerated shipping in the 1870s there was also a growing market for grass-fat beef from the Great Plains in the populous east and on the other side of the Atlantic.

The Turkey Track breeding herds proliferated. And the steers, grazing on prime pasture lands of the Cherokee Strip, put a thick covering of yellow fat over their ribs. The partner-owned operation flourished.

Things began to sour when a prolonged drought that started in the spring of 1892 seared the southern plains, and the cattle market turned downward. The final undoing for the Cresswell Cattle Company on the Cherokee Strip was the pressure brought to bear by the arrival of the Fort Worth and Denver and the Southern Kansas Railroads in Oklahoma and the Texas Panhandle. The railroads, wanting to capitalize on selling land the government had granted them along their right-of-ways, were doing a great deal of promoting back east, urging people to follow Horace Greeley's advice and "Go to the West."

The federal government had designated Oklahoma as Indian Territory and had granted land there to more than thirty displaced tribes. Among the transplanted tribes were the Cherokee, Creek,

Choctaw, Chickasaw, Seminole, Cheyenne, Arapahoe, Kiowa, Comanche, and Apache. Congress finally capitulated to lobbying by citizens and the railroads and voted to renege on treaties with tribes. The Indian Territory was now public domain. The Oklahoma Land Rush started in 1889. The largest and wildest rush took place on September 16, 1893 when the Cherokee Strip along the Canadian River in western Oklahoma was opened for settlement.

Once again, government edicts and the nester with his plow, his barking dog, and his damned barbed wire forced Cresswell and his partners to look for new horizons.

ROUNDUP

The Circle Diamond in Colorado became a combination cow-calf, steer, and horse-raising operation. Steer calves from their own breeding herd, together with purchased steers, were run on grass until they were three and four years old before they were trailed or railed to market. While deep snows and northers could be hard on steers, the losses were not nearly as devastating as with a breeding herd. The company was also running a large band of brood mares and stallions on their Colorado range. The colts were gelded and broke to the saddle by the cowboys as threes, each man being assigned several broncs each spring to add to his string of six to eight saddle horses. The best fillies were added to the broodmares and the rest were sold.

The story of the contribution of the horse and mule to the development of the Great Plains is a well-earned tribute.

> Of all God's creatures I endorse
> Most heartily the one called "horse."
> That on this creature man might sit
> No doubt is why God made him split!
>
> *Rawhide Rhymes,* by S. Omar Baker

The ranch headquarters for the Bloom Cattle Company in Colorado was at Thatcher, thirty-two miles northeast of Frank's home in Trinidad. From the base at Thatcher the operation ranged from Chicaguagua Creek and along the Purgatoire River on the east, north to the vicinity of the Santa Fe Railroad line between La Junta and Pueblo and west to the eastern ramparts of the Sangre de Cristo Mountains. Frank established cow camps at strategically located properties as bases for the men riding herd on the thousands of Bloom Cattle Company livestock which grazed this southeastern Colorado country.

When the seed heads of the prairie grasses were rolling under gentle breezes heralding harvest time for men of the range, the Circle Diamond cowboys would make ready to gather the cattle, work the herds, and ship the marketable steers. They would see to their string of saddle horses, making sure that each mount was sound and had a new set of Diamond horseshoes nailed to its hooves, for they would leave a million horse tracks on the vast Company spread during the roundups. Anton, the cook, would put his cooking gear in the chuck box—fashioned after Charlie Goodnight's—and provisions in the mess wagon to feed a crew of hungry cowboys for a couple of months.

Cattlemen who ran stock on the vast open range of southeastern Colorado would send their crews out with ranch wagons on a community roundup. The owners or managers of the different outfits agreed on a system whereby each outfit had a particular section to work. This ensured that they would not overlap or miss an area.

When working the range claimed by the Bloom Cattle Company, one or two outside riders, or reps as they were called, from the Prairie Cattle Company, the Reynolds Cattle Company, and the other outfits whose cattle might have strayed onto Bloom territory would ride with the Circle Diamond crew. A roundup crew might consist of some twenty cowboys, a cook, his swamper, and a horse wrangler who herded a remuda of maybe one hundred and fifty horses. They would spend from four to six weeks gathering, sorting, branding, and moving cattle.

The working of cattle on large ranges in the west has changed some, what with the introduction of barbed wire fencing, pickups carrying the chuckbox and pulling horse trailers, advances in the science of veterinary medicine, and smaller crews. But a cow still gestates for nine months and goes to water the same way, the fundamentals of working a herd of cattle are much the same today as they were eighty years ago.

The morning star was sparkling in the east and there was "frost on the pumpkin" when the roundup crew at Circle Diamond headquarters tossed their rolled bedding and war bags into the hooligan wagon. They mounted their top circle horses and rode out to start the work as the rays of the sun crept above the horizon. Anton gave a shrill whistle and slapped the reins across the rumps of the mess wagon team; his swamper, driving the hooligan wagon, hollered at his mules and the iron-rimmed wheels cut through the turf making tracks toward the designated site for the roundup.

For the work of gathering the cattle off Circle Diamond range, Frank's brother Wilbur would give the cowboys their powders, cowboy lingo for orders. The best hands would be sent to ride around the outside circle of the area to be gathered, pushing cattle they picked up inward toward the site selected by Wilbur for working the herd. The rest of the cowboys, riding singly in ever-tightening, concentric circles, would pick up cattle pushed their way, throw them together with any they found in their assigned area, and head them toward the center of the roundup. It called for teamwork. A clean gather depended on each man being in the right place at the right time, remaining alert, and being acquainted with the lay of the land and the nature of a bovine critter.

To work the big country, a cowboy needed to be a horseman, mounted on a hard-twisted horse with a lot of heart, especially in rough, brushy country. He had to have sharp eyes so as not to overlook any cattle, and have an instinct for spotting those wily ones that were easily spooked and would try to brush up or run off. He must make sure that every "wet" cow he picked up had her calf with her and keep the gathered cattle moving toward the roundup spot. He might jump a wild cow, a maverick, or pick up a bunch-quitter that wouldn't stay with the gather. Spurring his horse to catch up with the runaway he'd jerk down his lariat, flip out a loop, race up alongside the fleeing animal, and sail the loop over the bovine's shoulders and down to ensnare its two front feet. He'd jerk the slack in his rope to tighten the loop while reining his running horse to ride away from the renegade at a sharp angle. Its front feet would be jerked out from under it, and it would take a hard fall.

"Bust them wild ranahans a time or two an' they'll hunt the middle of the herd and damned sure stay there," an old-timer might have said. A roundup in those days was sure as hell no place for a greenhorn.

The mounted hands would drive the cattle to the place where Wilbur had told Anton to set up camp. The cowboys would hold the bunched cattle while the cows and calves paired and the herd settled down. Wilbur would send a few men at a time to the chuck wagon for dinner and to change horses. They would lope over to

Circle Diamond remuda, Thatcher, Colorado. Courtesy Trinidad Collection, Colorado Historical Society.

Springtime branding of calves. Diamond A Ranch near Wagon Mound, New Mexico, circa 1930.

(Left to right) Jim Loyd, Louis Hughes, Jack Connelly, Darwin Daniels, Charlie Green, branding on the Diamond A Ranch, circa 1930.

where the horse wrangler was holding the horse herd, pull their saddles off the sweat-encrusted backs of their circle horses, and turn them loose with the remuda. The weary horses would kneel to roll on their backs in the dust, then get up, stand spraddle-legged, and give themselves a vigorous shake. The bowlegged, tired, and hungry cowpunchers would head for the wagon.

Anton, working in the shade of the fly at the back of the mess wagon, would have waiting in iron skillets and Dutch ovens pan-fried steak, potatoes, gravy, sourdough biscuits, a dessert made from dried fruit, and gallons of coffee strong enough to float a horseshoe. And if they had just butchered a beef there would be son-of-a-bitch stew (diced liver, heart, tripe, kidney, tongue, and brains seasoned and boiled in a Dutch oven until a green scum covered the simmering concoction) as the specialty of the day. A man was allowed just time enough to squat cross-legged on the ground, wolf down his dinner, lean back against a wagon wheel or stretch out on the grass, and enjoy a hand-rolled Bull Durham before it was time to go back to work.

The cowboy, picking his teeth with a whittled match stick, would return to where he had laid his saddle beside the single-rope enclosure that held the horses. With bridle in hand he would call out to the roper which mount he wanted for the afternoon work. The man would flip a hoolihan loop—"no bigger than a Stetson hat"—backhanded across in front of him and over his head to sail out above the crowded remuda and settle around the head of the desired horse. The roper would snatch the slack to tighten the loop around the throat latch, then lead it from the bunch and turn it over to the cowboy.

Horsemen would drag in wood at the end of their lariats. On the open prairie they would use the supply Anton carried in a old hide slung under the belly of the hooligan wagon. The branding crew would dig a fire pit and start their fire. When there was a fire bed of red-hot coals under burning wood, the business-end of the branding irons would be put into the coals until the brand at the end of a long iron handle was white-hot. The younger hands that would be the flankers would pair up, take off their chaps and spurs and get ready for the grueling work of wrestling calves to the ground and holding them while the man with the knife and the brander did their work. It took teamwork, know-how, and stamina for the two flankers to wrestle a struggling calf to the ground and hold it in a viselike grip while the brander laid a white-hot iron on to burn a golden-brown mark on the hide, and the man with the knife did surgery to the ear and the scrotum. If the flanker on the rump didn't hold the hind legs tight the calf could jerk loose and kick the castrator in the head.

The man chosen to do the castrating and ear marking would hone his knife until it was razor sharp. One of the greenest of the crew would be told to handle the dope bucket. It would be his chore to run to each downed animal that could no longer boast the bagged appendages that marked it as a potential herd sire and daub the severed scrotum with a liberal smear of pine tar.

There was an exacting art to burning a brand on an animal. The iron had to be hot enough and applied with the right touch to leave a permanent mark of ownership on the hide. But not too hot and put on with a heavy hand, burning through the hide leaving a blotched scar instead of a readable means of identification. A "hair-brand" that only scorched the hair and did not sear the hide deep enough to leave a scar would disappear after a few months. Hair-branding was a trick of the trade for rustlers in the days of open range. They would put such a brand on a calf, then after the calf was weaned, pick it up as an unclaimed maverick, alter the ear mark, and put their brand on it.

The two men assigned to work in the herd would each pick their special mount for that job, saddle, mount, shake out their lariat and ride slowly in among the cattle. A full-eared, unbranded calf would be spotted. The roper would trail it until he had it paired up with its mama, flip a loop over the calf's head or around its heels and ride at a trot towards the branding fire with the bawling calf dragged along. The roper would call out the brand on the calf's mama to the branding crew. Calves might be dragged to the fire by the neck or the heels, but to make it easier on the men doing the flanking, yearlings were dragged in by the heels.

Older, unbranded mavericks were roped around the neck and dragged toward the fire. The second roper would fall in behind and pick up the animal's heels in his loop. The two men would then quickly turn their horses to pull against each other, drawing the lariats taut, stretching the maverick out to lay on the ground. The mounted men would hold the lariats tight, immobilizing the maverick while the men on foot did their part of the work.

Mavericks and dogies, that couldn't be paired with a cow for identity, were claimed by the ranch on whose range the roundup took place.

After the branding fire was doused and the irons were cleaned and set to cool, the work of shaping the herd would start. Part of the mounted crew would hold the gathered cattle while other cowboys would be ready to hold the various cuts; those animals that would be, for one reason or another, hazed away from the main herd. The cow foreman would ride his top cow horse among the cattle looking for marketable steers and other cuts belonging to the Bloom Cattle Company. Reps for the neighboring ranches would prowl through the herd looking for animals wearing the brand they rode for. Once a cut was sighted the rider would ease his horse in behind the animal and haze it to the perimeter of the herd. If it tried to turn back, the cutter's mount, with its natural cow-working ability, would whirl and charge to stay between the cut and the main body of the herd. When the cutter had the animal under his control and headed away from the herd, a nearby cowboy helping to hold the herd would slip in behind the cut and start it toward the group to which it belonged.

Each of the cowboys' jobs required know-how. A man working in the herd had to be well mounted on a quiet-dispositioned horse with an inherent ability to out-think the bovine. The man had to know cattle, think like a cow, be able to read and know the brands and ear marks of the different ranches, and do his work in the herd without causing a spill.

A thousand churning hooves in a herd being worked would cut through the turf and into the top soil to raise a cloud of dust that irritated the eyes and clogged the nostrils. The smoke and acrid smell of burning hair and hide under the branding iron added body and pungency to the atmosphere. The pace was always hurried, balancing the work to be done against the approach of the setting sun. Sweat stained shirts and ran from under hat brims. Cows bawled, shook their heads, slavered, and threatened the men bending over their bellering calves.

After the branding and sorting was all done, the different herds would be tallied. A man who could count cattle accurately was held in the highest esteem among cattlemen and cowboys. The herd to be counted would be strung out, cowboys at the back and on the sides keeping the cattle moving forward, shaping the herd like a funnel. If there was only one class of cattle to be counted, the counter would sit his horse beside the neck of the funnel, taking a tally as they streamed by him. If there were more classes to be tallied than a man could reasonably be expected to keep track of, a second counter would be on the opposite side making his count. For a large herd, the counter would break his tally at a hundred, registering each hundred by tying an overhand knot in his bridle rein or transferring a rock from one hand to the other. Herds comprised of hundreds of cows, bulls, grown steers, and calves were bought and sold on the basis of such a count.

The Circle Diamond bulls, cows and their calves, heifers, and steers too young to market were counted and turned loose to scatter on the company's range. The marketable steers were trailed to Thatcher to be loaded on Atchinson, Topeka & Santa Fe Railroad stock cars and sent to Dodge or Kansas City. Cattle belonging to neighboring ranches would be driven by the ranches' representatives to their home ranges.

Wilbur would send a supply wagon from headquarters to replenish Anton's mess wagon and bring the cowboys a fresh supply of tobacco and cigarette papers. The outfit would move to another part of the country and start working another herd.

THE CAPITALISTS

Although he was cut from a different bolt of cloth than most of the men of his day who invested heavily in cattle, the self-made John Wesley Iliff was known as the Cattle King of Colorado. Iliff was one of the first of the large-scale entrepreneurs who would turn their herds of cattle numbering in the tens of thousands out to graze among the bleaching bones of the bison along the South Platte River. Iliff died in 1878 at the age of forty-eight. He dedicated a large part of his estate to the founding of the Iliff School of Theology in Denver. Frank Bloom's daughter, Alberta, married John Iliff's son in 1897. The Trinidad *Chronicle-News* on July 7, 1897 described the groom: "William S. Iliff is one of the leading businessmen of Denver, universally respected, and perhaps the richest man of his age in Colorado."

Owen Wister, who wrote the classic novel *The Virginian,* sent journals relating his experiences on the western frontier back to publishers in the east where they were widely read and discussed. Writing from the Wyoming Territory, he described the Cheyenne Club as "the pearl of the prairies." This private club was built in 1880 by and for cattlemen. It was a three-storied structure with hardwood floors and plush carpeting, sitting like an exclusive country club, not back east beside a manicured golf course but on the Front Range out in the middle of the prairie. Membership was limited to two hundred captains of the cattle industry. The club offered lodging, elaborate dining facilities where guests dressed for lunch and dinner, a basement vault stocked with imported wines, two grand staircases, a spacious lobby, a smoking room, and a reading room.[1]

In May 1882, the price for beef cattle on the Chicago market was the best it had been since 1870. Investment dollars and pounds sterling, earmarked for cattle, poured into the west.

From 1882 to 1886 there were 176 cattle companies incorporated in Colorado with a capitalization of $74,296,000; in New Mexico there were 104 companies capitalized at $23,935,000; Wyoming listed ninety-three companies capitalized at $19,509,000.[2]

Granville Stuart, miner and pioneer cattleman in Montana, reflected in later years:

> It would be impossible to make people not present on the Montana cattle ranges realize the rapid changes that took place on those ranges in two years. In 1880, the country was practically uninhabited. One could travel for miles without seeing so much as a trapper's bivouac. Thousands of buffalo darkened the rolling plains. There were deer, elk, wolves and coyotes on every hill and in every ravine and thicket. In the whole territory of Wyoming there were but 250,000 head of cattle, including dairy cattle and work oxen.
>
> In the fall of 1883, there was not a buffalo remaining on the range, and the antelope, elk and deer were indeed scarce. In 1880, no one heard of a cowboy in this "niche of the woods' and Charlie Russell had made no pictures of them; but in the fall of 1883, there were 600,000 head of cattle on the range.[3]

To the boardroom entrepreneurs and the young scions of wealthy families investing in frontier ranching, bigger was better. One of the most extensive range-cattle operations in the history of the cattle industry in the United States was the Prairie Cattle Company, Ltd. Their involvement was instigated by the brokerage firm of Underwood, Clark, and Company out of Kansas City. They convinced capitalists in Edinburgh and London that there was a golden harvest to be reaped with the herbivorous bovine grazing on the open range in the "outback of the Colony."

The purchase of the Jones brothers' JJ properties in southeastern Colorado in 1881 was the beginning. From his office in Trinidad, Colorado, Murdo Mackenzie, general manager of the Prairie Cattle Company, Ltd. from 1885 to 1890, continued expanding their cattle operation. Their cattle and horses ranged on the Arkansas Division in Colorado, the Cimarron Division in northeastern New Mexico and western Oklahoma, and the Bar CC Ranch of the Canadian Division in the Texas Panhandle. The three divisions totaled a land mass of over 5.5 million acres, a little shy of eighty-six hundred square miles, most of it public domain. The company, at its peak, carried an annual inventory of some 139,000 head of cattle on its books.

Catching the scent of money on the westerly winds, other investors were quick to follow Iliff, Goodnight, Prairie Cattle Company, and the Thatcher brothers and their associates into the cattle business. The Swan Land and Cattle Company, Ltd.'s 1883 prospectus to potential investors stated: "The business of cattle-raising in the Western States of America is now acknowledged to be highly remunerative." The directors of the company backed up their statement by acquiring some thirty thousand acres of land, part in full title and part in process of title, with range rights covering a vast spread of public domain. In addition, the company purchased over 860 sections granted to the Union Pacific by the government, and since these 640-acre tracts were checkerboarded with public domain, the company actually took control of over more than a million acres. In all, the Swan Land and Cattle Company, Ltd. controlled over 3.25 million acres in 1884.[4]

In 1885, the Capitol Freehold Land and Investment Company, Ltd. organized and incorporated the 3,050,000-acre XIT Ranch on lands patented by Texas in the Panhandle. As many as 150 cowboys, drawing their mounts from some one thousand head of saddle horses, rode herd on 150,000 head of cattle and burnt the XIT on the right ribs of some thirty-five thousand calves each year.[5] Starting in the late 1880s the XIT cowboys would gather the cattle in the fall. They would cut from the herds the yearling steers, throw them together, and start them on the thousand-mile trail to the company's range in Montana. There the steers were turned loose to graze on a vast spread that lay between the Yellowstone and the Missouri Rivers. These Texas-raised cattle grazed on a thousand hills where the heads of the thickly-sodded prairie grasses brushed their dewlaps. They watered in the creeks, in the coulees, and in the old buffalo wallows after summer rains; and shaded up in the peachleaf willow thickets to chew their cud in the heat of the day.

By 1886, Charlie Goodnight and his financial backer, the Englishman John Adair, had expanded their JA Ranch in the Palo Duro Canyon country to encompass something more than 1.35 million acres, much of it patented or leased from the state of Texas. The ranch supported over a hundred thousand head of cattle year round.[6]

Under Goodnight's innovative guidance, the JA Ranch was a leader in range and ranch management and in herd development.

Weaning Hereford calves about 1930, Diamond A headquarters.

The ranch was fenced and cross-fenced for herd and grazing control. Early on, Goodnight introduced blooded bulls and female stock to upgrade the quality of the original longhorns. He was one of the leaders in establishing range management and the introduction of Hereford cattle, which was to become the dominant breed in southwestern ranching.

Dutch investors had purchased the Maxwell Land Grant, a grant from the Republic of México, which lay mainly in north-central New Mexico. A conflict of ownership between the corporation and preempting settlers over grant boundaries ignited the Stonewall War in 1888. The confrontation pitted the settlers who claimed preemption rights against the political and economical might of the Dutch corporation's Maxwell Land Grant Company and its hired gunmen. The corporation's legal battery claimed the area under dispute, several hundred thousand acres of pristine agricultural land, by virtue of a current survey which established the controversial northeastern cornerstone not to be on Raton Peak overlooking Raton, New Mexico, as was supposedly set by the original grant document, but north some ten miles on Fisher Peak overlooking Trinidad.

The settlers claimed the Maxwell Land Grant Company had paid to have the cornerstone moved. The company won the argument in the courts, and in total claimed some two million acres patented by the validated grant in New Mexico and Colorado.[7]

Some directors of the Santa Fe Railroad, the Seligman brothers—New York bankers—and a group of Texas cattlemen organized the Aztec Land and Cattle Company in 1885. Cowboys and cattlemen called it the Hash Knife after the brand (⊤). The company purchased from the financially-strapped Atlantic and Pacific (A.&P.) Railroad 1,059,560 acres for $529,500. This was railroad grant land between Flagstaff and Holbrook, Arizona. The government had granted forty sections—25,600 acres— to the A.&P. for every mile of track completed. The granted sections straddled the right-of-way and were checkerboarded with public domain. The Hash Knife (or Hashknife) thereby controlled 2,119,120 acres. To maintain possession the company had a crew of gun-toting toughs hired to run off any settlers or ranchers who had ideas of claiming any of the public domain that lay between the granted sections or that abutted the Hash Knife.[8] The backers of the Aztec Land and Cattle Company, "didn't want to claim it all, just what joined 'em."

The Matador Land and Cattle Company, Ltd., had its accounting office in Dundee, Scotland. In the 1880s, the Matador empire embraced patented lands and range rights in Texas, Montana, and the Dakotas. In Colorado, the Arkansas Valley Land and Cattle Company, Ltd., a London-based firm, ranged their cattle branded S S on some two and a half million acres in the southeastern part of the state.[9] The Rio Grande Land and Cattle Company had an authorized capitalization of ten million dollars, while the Antonito Land and Cattle Company was capitalized for a paltry five million dollars.[10]

The transition brought on by the cattle boom across the Great Plains rivaled, in regional impact, the Industrial Revolution in the East. The killing winters in 1885, 1886, and 1887; the collapse of the cattle market in 1888; and overstocked ranges brought cattle investors face-to-face with the harsh reality of the perils and volatility of the cattle industry on the Great Plains. The three compounding disasters caused the financial ruin of many capitalists and ended the cattle boom.

In 1895, only fifty-two percent of the major cattle companies that had stocked the midcontinent in the 1880s were still in business. By 1900, only thirty-one percent had survived. Big corporations and small ranchers alike fell on hard times; many turned belly-up. The bigger they were the harder they fell. Many absentee-investors paid for the cattle they purchased, not on the basis of an actual count but on the seller's stated inventory. In the vast open-range country, taking an accurate count of igratory cattle was almost impossible. It took a company running thousands of cattle, under the direction of a capable and trustworthy ranch manager, several years to ascertain an approximate inventory of its cattle herd. Very often, a range tally and an accounting on the corporate books would disclose an alarming discrepancy. To compound the discrepancy, absentee owners assumed each cow would produce a calf per year; in truth, only about sixty-five percent of the cows calved.

The Swan Land and Cattle Company went into the hands of a receiver after writing off two hundred bulls, twenty-two hundred steers, and fifty-five hundred cows.[11] The Cheyenne-based Union Cattle Company's books showed fifty-five thousand head of cattle valued at sixty dollars a head on the books in 1883. A roundup and count of the cattle in the spring of 1887 tallied something less than thirty thousand head, and the market had fallen to twenty dollars a head. In the fall of 1886, the Montana-Nebraska based Niobrara Land and Cattle Company claimed to creditors and investors that they had thirty-nine thousand head of cattle valued at over a million dollars. A year later they could only account for nine thousand head and their assets fell to a quarter of a million, on which they claimed they would be fortunate "to garner fifty percent." Their liabilities stood at $350,000.[12]

Almost any individual or corporation that had an investment in cattle on pasture from the hundredth meridian to the crest of the Rocky Mountains and from Canada to Mexico in the 1880s was subject to adverse weather and poor markets. However, not all fared the same. A larger percent of the big operations went under than did the smaller ones. Most of the smaller operations were under the direct supervision of the investor, who had experience in the field of cattle husbandry and kept a tight rein on the management. The field operation for most corporations was under the management of an employee. Reports from the field and the profit and loss statements were reviewed by the directors sitting in a distant office.

Management by a salaried employee who was either not qualified for the responsibility or whose area of responsibility was too large to be supervised closely, compounded by initial purchases of

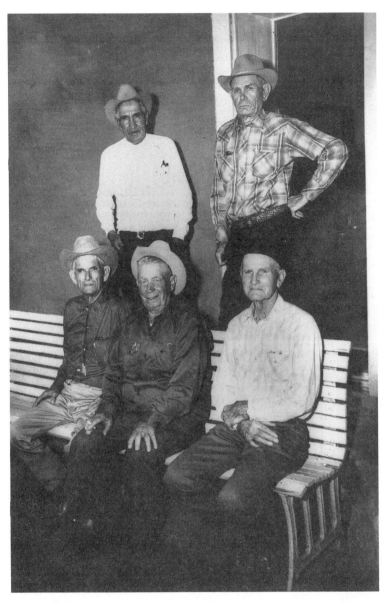

Diamond A cowboys at a company reunion. (Left to right, standing) Albert Lopez, Tony Maipen. (Seated) Tom Russell, Kitch Lavender, Joe Lopez. Trinidad Collection, A. R. Mitchell, photographer. Courtesy Colorado Historical Society.

cattle inventories that were overstated, by killing winter weather, and by a prolonged depressed market proved to be a backbreaking burden for many of the large companies.

But some of the corporate giants managed to survive. The Prairie Cattle Company weathered the storm, as did the Aztec Land and Cattle Company. The Matadors, Charlie Goodnight's JA Ranch, and the XIT rode on into the 1890s. The Thatchers and their associates, with farsighted directives and financial supervision from the Thatchers and hands-on management by Bloom and Cresswell, suffered losses, learned from the experience, recouped, and continued to strengthen and expand their holdings from their base in Pueblo.

RANCH ON THE RIO HONDO

When Frank established the Bloom Cattle Company's Circle Diamond Ranch in the Territory of New Mexico early in 1885, he selected a site for the deeded base unit in the Hondo Valley some forty-five miles west of the growing village of Roswell and the Pecos River. The ranch headquarters for the operation was near the post office at Picacho. Seven miles up the valley west of Picacho, the Rio Bonito and the Rio Ruidoso came tumbling out of the rocky, forested foothills of White Mountain to form the Rio Hondo. Lincoln, the county seat, was on the south bank of Rio Bonito, twelve miles northwest of the junction.

The Hondo Valley widened below the foothills of White Mountain as it coursed through the rolling hills of grassland on its way to the Pecos Valley. The Rio Hondo nurtured dense stands of bunch grass, sinuous vines tangled in the underbrush, and gnarled cottonwoods with high water marks on their trunks higher than a man horseback could reach; evidence that the gurgling creek could turn ugly and deadly when it was swollen by flood waters that came rampaging down the valley.

By the fall of 1886, Circle Diamond headquarters on the Hondo consisted of a large adobe home, a bunkhouse, cook shack, barns, saddle shed, corrals, and several small wire-enclosed traps used to hold the night horses, the milk stock, and any range cattle that were to be held over for a day or two. Log cribs filled with rocks dammed the Rio Hondo at intervals to divert water into ditches that led to cultivated fields in the deep alluvium along the bottom of the valley. Alfalfa, corn, and oats were the main crops grown.[1]

John and Mahlon Thatcher, as the financial backers and president and secretary-treasurer respectively of the Anderson Cattle Company, were finding that William Anderson, as the general manager, was a bit too headstrong to fit with the teamwork employed by the Thatchers and their associates. He was also proving to be rather unscrupulous and heavy-handed when it came to dealing with his Mexican neighbors, many of whom had lived in the area for generations and lacked sophistication in business negotiations. With the financial backing of the First National Bank of Pueblo and in the name of the Anderson Company, Anderson bought 320 acres of land in the valley from Felipe Chaves de Candelario for the sum of thirty dollars, the going price of an ordinary saddle. While this was to the company's advantage, it made for poor public relations in an area populated primarily by the natives, who tended to be hostile toward arrogant, high-handed gringos.

Even his Anglo neighbors found Anderson to be offensive and overly aggressive in his push to acquire land. A letter dated April 15, 1885:

A. J. Stewart, Esq.

Lookout P. O. N. M.

Dear Sir:

 I see you have a filing on the Rio Hondo in my Range, which you filed over a year ago. I learn that you have secured a range elsewhere and suppose you made this filing before you had a name. You would confer a great favor on me if you will relinquish your filing on this piece on the Hondo as I have secured the water rights on the Hondo from your Timber Culture down for 15 miles and nearly all the water rights for 10 miles above you. Your Timber Culture is in a condition to be contested at any time and supposing you will have no use for the place now I trust you will relinquish it at once so that I may go to work right away and secure it before someone contests your filing and locates on my range as "jumpers" are very numerous in these parts.

Yours resp.

W. E. Anderson

[Author's note: Timber Culture was a patent under the Homestead Act.]

At a board meeting of the Bloom Cattle Company it was decided that a change in the handling of their business venture in Lincoln County would be to the best interest of the company. Early in 1888, the Thatchers took over the Anderson Cattle Company. They closed it out and incorporated the assets into the Bloom Cattle Company. This added hundreds of acres of deeded land interspersed among a scattering of other settlers along the Hondo Valley east of the Circle Diamond headquarters at Picacho. The company also acquired the appertaining water rights from the Hondo River and range rights to thousands of acres that had been a part of the Anderson Company on the areas in the hills north and south of the deeded lands. The board members of the Bloom Cattle Company elected to call this addition to the company the Diamond A Ranch. They bought out Anderson's personal property, which included his cattle, some horses and the Diamond A brand.

At about this time Frank negotiated to buy the Bar H Ranch from James Sutherland and Jimmie Farrel.[2] The deeded land of the Bar H was about five miles down the valley from Circle Diamond headquarters at Picacho, lying between Circle Diamond patented land and that of the Diamond A. James Sutherland was the son of Jack Sutherland, the man who had bought the Adobe Walls ranch in about 1877 and had been a neighbor to the Cresswell Land and Cattle Company.[3] Jack Sutherland was William Anderson's stepfather. Anderson had hired his stepbrother, James, as the cow boss for the Anderson Cattle Company.

As the manager for the combined Circle Diamond and Diamond A properties, after the departure of William Anderson, Frank Bloom found James Sutherland, a short, heavyset, likable fellow, to be a company man, a man who possessed a good eye for the qualities of a horse, and an experienced cowman. He was considered a good neighbor by the settlers in the valley. Frank promoted him to foreman of the combined ranching operations.

Homesteaded and preempted acreages in Lincoln County sprouted along live-water streams, the Pecos River, Rio Hondo, Rio Peñasco, and Rio Felix. Settlers also filed for homesteads and squatted around the natural springs and along intermittent tributaries like Seven Rivers and Tres Ritos, creek beds that generally had potholes of water even during the dry season.

With the exception of large areas of rangeland without dependable sources of water for livestock, before the common employment of deep drilling rigs and the Eclipse windmill, Lincoln County was a cattleman's Garden of Eden. But even Eden was not without problems. Ranchers had to contend with the whimsical weather of the desert southwest, rustlers, predators, the flesh-

Sheep running under the watchful eye of a herder and his dog on the Diamond A Ranch, circa 1926.

eating screwworm, and Apaches with an inclination to eat the white-eyes' cattle that strayed too close to their reservation.

But there were redeeming features. First and foremost for stockmen, and especially those who experience ranching farther north, were the mild winters. Then there were the broad valleys with heavy stands of grasses and forbs and hillsides sodded with a blanket of short grasses which put a thick layer of backfat on steers, heavy wool coats on sheep, and full bags on cows and ewes. Along the creeks and rivers the valleys afforded bottom land for irrigated cultivation.

The decision to move part of the Circle Diamond breeding herd from their range in southeastern Colorado to the Hondo Valley in the summer of 1885 proved to be a prudent one. It put those cattle beyond the reach of the disastrous winters of 1885–86 and 1886–87 when an estimated twenty-five to forty percent of the cattle running in southeastern Colorado froze to death or died of starvation.

EXPANDING HORIZONS

Although their financial and personal interests in ranching were extensive, John and Mahlon's primary business was banking. The First National Bank of Pueblo had grown from $176,004.30, as shown in the 1871 financial statement, to $2,442,334.11 in October 1890.[1]

As finance is the lifeblood of expansionism, the Thatchers' financial institutions followed pioneering settlers and miners, aiding in the development of Colorado. The First National Bank of Pueblo opened a branch in Silverton in 1883 after silver was discovered in the San Juan Mountains. Branches were also opened to accommodate the budding farming areas in Lamar, Rocky Ford, and Florence. The First National Bank of Pueblo was the parent bank of the Bent County Bank of Las Animas, the American National Bank of Alamosa, the Miners and Merchants Bank in Ouray, the Montrose National Bank, and the Capitol City Bank in Denver.[2]

The Thatchers had assisted in financing the opening of the First National Bank in Trinidad in 1875. In 1886, they bought out the other major stockholders. Mahlon was elected president and Frank Bloom vice president.

In 1890, John and Mahlon moved their Pueblo banking business into new quarters at Fourth and Main. It occupied the entire first floor of the three-storied Grand Opera House which was built on land donated by John Thatcher and Perry Baxter. The main entrances, off Main and Fourth, opened into small vestibules whose walls, ceilings, and floors were of polished marble. The spacious lobby was tiled with marble. The large vault had walls of steel and was equipped with an automatic, innovative time lock.

The only mining activity of any significance that the Thatchers ever undertook, aside from a coal mining venture with Frank in Las Animas County, was financing the Revenue Tunnel at Ouray, Colorado. This tunnel made possible the opening of the Camp Bird Mine which produced millions of dollars worth of silver.

The directors of the Bloom Cattle Company understood the implications of President Cleveland's edict disallowing fencing of public lands by private interests. This edict portended the ending of the era when Circle Diamond cattle would graze on free grass in Las Animas, Pueblo, Huerfano, Bent, Prowers, and Otero Counties.

The fact that they were looking for more stable pasturage for Circle Diamond cattle is substantiated by a letter written by Frank Bloom in the withering heat of the droughty summer of 1888 from Greensburg, Kansas regarding negotiations on leasing a 150,000 acre ranch for the Bloom Cattle Company. In the letter, written on Queen City Hotel stationery, he described the country and the ranch: ". . . as good grass and fat cattle as I have seen for some years. They have a fine ranch and as well watered as I have ever seen. We will put in 5,000 steers."

After the Bloom Cattle Company took over the Anderson Cattle Company it was decided that they would thereafter brand horses on the left shoulder with Circle Diamond and breeding cattle on the left rib with the Diamond A. All the steers belonging to the Bloom Cattle Company would be branded on the right ribs and hip with the Circle Diamond. The ranch operation in Colorado would be the Circle Diamond and in Lincoln County, the Diamond A.

New Mexico had yet to see the influx of immigrants that were following the railroads into Colorado and Texas. Title to the bulk of the land of the territory was still held by the federal government. The Bloom Cattle Company livestock grazed on open range, north from the main ranch in the Hondo Valley to somewhere south of Fort Sumner and south to the Seven Rivers country, depending on rains and available stock water. To the west lay the natural barrier of the rugged Carrizozo and Sacramento Mountains, and to the east the plains that sloped to the silt-laden, alkaline waters of the Pecos River. This was an area approaching seven thousand square miles, or nearly four and a half million acres, which consisted mainly of public domain. Unlike the plains of eastern Colorado, the rocky, semiarid hills that lay east of the foothills in Lincoln County were ill suited for the nester's plow. But they were blanketed with perennial grasses that provided

Range branding scene, circa 1900–02, Circle Diamond Cattle Company (Bloom Land and Cattle Company). Courtesy Montana Historical Society, Helena.

excellent pasturage for cattle, sheep and horses. The limiting factor to livestock on this spread of open range was stock water.

In the summer of 1887, the Diamond A wagon and a crew of cowboys threw in with twenty-one wagons representing other ranches that grazed their cattle on the open range along the Pecos Valley from Fort Sumner to Seven Rivers. They gathered for a roundup on the vast country that had once been claimed as the range rights for "Uncle" John Chisum's some 60,000 head of cattle. "The Cattle King of The Pecos" had died in 1884.

Two hundred and fifty cowboys, with a remuda of over two thousand head of horses and mules, worked at gathering strays that had drifted south with the winter storms of 1886–87. It took over a month to gather and work thousands of head of cattle from the rangeland from as far north as the Canadian River to the Valley of Seven Rivers, near the Carlsbad Caverns.

Along the mostly deeded land flanking the Hondo River, Frank had Sutherland construct ditch systems that carried water from the river, allowing them to irrigate nearly a thousand acres of fertile bottom land. The sparkling waters in the Rio Hondo came from numerous springs and snowmelt from the alpine mountain range to the west. June was normally the hottest and driest month of the summer. July and August rains almost always freshened the little Hondo River to supply irrigation water.

A young fellow by the name of John Survant had wandered into the Hondo Valley in 1885 looking for a job. James Sutherland gave him a cowboying job on the Bar H Ranch. Survant was from Missouri. He had graduated from Chillicothe High School in 1882 at the age of eighteen.[3] When the Bloom Cattle Company added the Bar H to its Hondo Valley holdings, John Survant hired on as a Diamond A 'puncher.

The young man impressed Frank Bloom. He was a clean-cut fellow, a cut above the average, drifting cowboy, and one who did not use profanity; this was an additional recommendation for Survant, as Frank was a deacon of his church in Trinidad. Always on the lookout for men with promise, Frank transferred Survant to the Circle Diamond, where he would report to Wilbur Bloom.

✳

The dreams and hopes, spawned by the Ghost Dance, of a miraculous resurrection of their past and the demise of the white

John Survant, 1913. Courtesy Montana Historical Society, Helena.

man were buried with the frozen bodies of Sioux and Northern Cheyenne men, women and children massacred near Wounded Knee Creek in South Dakota in 1890.

The Seventh Cavalry had extracted grim retribution for its total defeat at the hands of the Oglala Sioux and the Northern Cheyennes on the Little Big Horn fourteen years earlier. The surviving Sioux and Northern Cheyenne were loaded on boats and

Noon camp, Circle Diamond Roundup wagon, circa 1900.

Beef herd on road to market 1900, Bloom Cattle Company, Circle Diamond
Ranch. Courtesy Montana Historical Society, Helena.

Group of cowboys, Malta, Montana, 1901, probably of Circle Diamond Ranch. Left to right: Jimmy Stowe, Letch Lemon, Bill Lyons, Bill Mason, Buck Hardin, Henry Plott, Spence Henry. Courtesy Montana Historical Society, Helena.

Part of Circle Diamond herd on water at Freewater Crossing on Milk River, circa 1902. Courtesy Montana Historical Society, Helena.

Circle Diamond Ranch near Malta, Montana, 1903. Courtesy Montana Historical Society, Helena.

shipped down the Missouri to reservations in the Dakotas. The vast grassland of the northern territories of the Great Plains lay silent.

In a meeting of Bloom Cattle Company officers in 1891 it was decided that, in view of the shrinking open range and the competition with other cattlemen and farmers in southeastern Colorado, the company should explore regions farther north. Bloom was instructed to travel to northeast Montana and check out the feasibility of establishing a ranch in the rolling grasslands north of the Missouri, one of the last expanses of sparsely-settled public domain available to cattlemen in the Great Plains. John Survant would accompany him.

The country north of the Missouri River in Montana was a late bloomer when it came to supporting great herds of cattle on its prairie. The trail herds coming north from Texas were the spawning of vast ranching enterprises in the southern region of the Great Plains. Pioneers like Goodnight had only pushed their herds as far north as it took to find good pasturage. The more distant, colder region was the last to be utilized. Some early cattlemen of the Montana Territory, like Conrad Kohrs, who started by selling beef to hungry miners and Granville Stuart, who was an early-day prospector, pushed open the pasture gate to cattlemen who were being crowded off public lands in the south. The tepee rings of the Crow, Blackfoot, and Sioux were only remnant circles of rocks and the bleaching bones of the Republican herd of bison lay encircled by vigorous emerald-green grasses that spotted the eastern plains of Montana.

The Bloom Cattle Company was a pioneer in developing the livestock industry in northern Montana. In 1891, Frank Bloom purchased land at the junction of Assiniboine Creek and Milk River, six miles north of the village of Malta, where he established a ranch headquarters for the Circle Diamond. Bloom, with John Survant at his side, hired a band of Cree Indians to put up a set of log ranch buildings and pole corrals. That fall, when the work was well along, Bloom and Survant rolled their beds and returned to Colorado.[4]

Survant returned to Malta early in the spring of 1892 to supervise the completion of the buildings and corrals, while Bloom was busy in Colorado putting together cattle for the new range. Twenty-five hundred head of steers were loaded in railroad stock cars at Thatcher. They were freighted to Orin, Wyoming, then trailed to Malta.[5]

It was the company's plan to stock the range with two-year-old steers, double-winter them, and market grass-fats. Bloom recruited cowboys off the Circle Diamond in Colorado to accompany that first herd north. When the cattle reached the end of the rail line, Bloom sought out a good man who knew the country and could point the herd to its new range. He hired a native of the region, Charley Stuart, the twenty-two-year-old son of Granville Stuart and his Shoshone wife, Aubony.[6]

Floyd Hardin's father was a wagon boss for the Circle Diamond in Montana and Floyd rode for the brand. In his book *Campfires and Cowchips,* Floyd charts the probable route Charley Stuart took in 1892 to pilot that first herd of Circle Diamond steers.

> Information on the exact route of this herd is vague, but the logical one seems to be up the North Platte River to the "Old Hog Ranch," then out across the prairie toward the Montana line. This route took them across many creeks and rivers, the Belle Fourche, Little Powder, Powder, and Yellowstone Rivers were crossed. From the Yellowstone, the route took them out past Sheep Mountain to a crossing on the Missouri somewhere near the mouth of the Big Dry, then westward between the Missouri and Milk Rivers to their destination, the new ranch on Milk River.[7]

After the cowboys had the Circle Diamond cattle located, Frank appointed Survant resident manager and told him to hire a crew, so Frank and his boys could backtrack the fifteen hundred miles home.

On September 30, 1895, a meeting of those who held the majority of shares of the capital stock of the Bloom Cattle Company, a corporation organized and existing under the law of the state of Colorado, met in Pueblo to amend the Certificate of Incorporation of said company. It was moved and seconded that the Articles of Incorporation be amended by adding: "And it is likewise the purpose of the Company to carry on and conduct part of its business in the state of Montana." It was signed by John A. Thatcher, president and Mahlon D. Thatcher, secretary.[8]

Within a few years, under the management team of Bloom and Survant, the Circle Diamond Ranch, as it was known by the locals, was pasturing between thirty and forty thousand head of cattle on range rights running east to west from Hinsdale to Chinook

Circle Diamond cattle near Hinsdale, Montana. Charles E. Morris, Chinook, Montana, photographer. Courtesy Montana Historical Society, Helena.

in the valley of the Milk River and south to north from the Milk River to the Canadian line. This was an area measuring roughly ninety-seven miles along the Milk River by something like thirty-seven miles to the Canadian border, or about two and a quarter million acres. *The Herd Quitter,* by Charles M. Russell, on display in the Mackay Collection in Helena, Montana, depicts three cowboys throwing their loops at a steer with a Circle Diamond brand on its ribs. It may be that Russell was, at one time, an employee of the Thatchers, drawing Circle Diamond pay.

CRESSWELL CATTLE COMPANY

The time had past when a rancher could preempt and put a fence around the perimeter of federal land he claimed under the livestock growers' ad hoc declaration of range rights. Settlers were spreading across the Great Plains challenging the stockmen's rights to public domain through legal avenues. Immigrants from beyond the hundredth meridian advanced, then retreated, and then advanced again, depending on the individual's grit and industry. They filed on tracts with natural waterings or where water could be pumped from shallow wells by a hand windlass or a creaking Eclipse windmill. Their stacked sod-block dwellings and their dugouts spotted the prairie. Their turning plows upended the prairie grasses, turned under buffalo bones and arrow heads, and turned up ancient burial grounds. Cattle and sheep belonging to a growing number of small ranchers were making inroads onto range rights claimed by pioneers.

A recanting of the treaties in 1889 between the governing body of the United States and various tribes, which had created the Indian Territory in what is today the state of Oklahoma, not only usurped the Native Americans' rights to the some two million acres set aside for reservations, but affected cattlemen who had entered into grazing contracts on the reservations with the Department of Interior and the different tribes. Realizing that land-hungry immigrants would soon be swarming over the Cherokee Strip, Henry Cresswell and Tony Day in 1891 set out to scout for new range.

Cresswell and Day very likely rented horses from a livery in South Dakota, maybe in the little settlement of Bronte on the bank of the Missouri River. They scouted west onto the Standing Rock Sioux Reservation, moving along over the rolling hills, taking note of the thickly sodded buffalo grass and gramas. They saw western wheat and wolf tail that would green up early in the spring and the turkey-footed tasseled heads of big and little blue-stem that would stand above the winter snows. They may have laid their camps in the shelter of draws where run-off waters and rivulets from springs fed currant bushes, gooseberries, snow-berries, grapes, wild plums, and browse. Draws would offer pro-tection for cattle from northers. The many bowl-like depressions, buffalo wallows, that dotted the prairie would hold water for Tur-key Track cattle and help scatter them over the vast spread. After prowling for days they must have agreed that here was the elbow room they needed.

In 1893, the partnership of John and Mahlon Thatcher, Henry Cresswell, Perry Baxter, and Tony Day agreed to close out their range operation in the Oklahoma Panhandle. They negotiated with the U.S. Department of the Interior and the E 6 Cattle Com-pany, a Canadian concern, to buy the E 6's cattle and take over its lease along the Grand River in the Standing Rock Indian Reserva-tion.[1] The company's main office would be located in the First National Bank Building at the corner of Fourth and Main Street in Pueblo, Colorado, and John would be president. There is no record of what the lease cost, but a few years later leases handled by the Bureau of Indian Affairs on the neighboring Cheyenne River Indian Reservation were going to non-Indian ranchers for three cents an acre per annum.[2]

In 1895, Day led a trail herd of thirty-six hundred head of Tur-key Track cattle across six hundred miles of prairie to the Cresswell Cattle Company's new lease.[3] They left the Oklahoma Panhandle, probably trailing along the old Chisholm Trail to-ward Dodge City, covering ten to twelve miles on a good day. They hazed the cattle into the Arkansas River at a crossing some-where between Cimarron and Dodge then drifted the herd due north. Day, more than likely, rode ahead to scout out a way to cir-cumvent the settlers' farms, as cattle coming from the south were unwelcome, suspected of carrying the parasite that caused "Texas fever." They crossed the North Platte and the Niobrara in central Nebraska, moving along east of the Black Hills and Wounded Knee Creek in South Dakota and crossed the Cheyenne River. Tony probably had the cook drive the chuck wagon into nearby Eagle Butte to restock the larder. The herd trailed across the Cheyenne River Indian Reservation to enter the Standing Rock Reservation somewhere near Trail City.

＊

Charles Goodnight and his partners John and Cornelia Adair had, in 1885, bought and leased 640 sections along the North Pease River in Briscoe, Hall, Motley, and Floyd counties in the Texas Panhandle. They named the spread the Quitaque Ranch, but it was known as the Lazy F outfit after the brand ($\diagup\!\!\curlywedge$). In 1887, the then widowed Mrs. Adair traded her deeded and leased rights to the Quitaque Ranch and the Lazy F cattle to Goodnight for his one-third interest in the JA Ranch and cattle.[4]

Goodnight, foreseeing the mass migration of settlers as the railroad laid rails across the Panhandle, sold his interest in the Quitaque Ranch to L. R. Moore of Kansas City, Missouri in 1890. Moore, in turn, sold the ranch and cattle to Frank Howard of New Orleans in 1895.[5]

Cresswell and Day expanded the Cresswell Cattle Company's business to include a lease from Howard on the Lazy F Ranch.[6] They bought some ten thousand head of Lazy F cattle, "one of the best herds in the country," according to Tony Day. They operated the Quitaque Ranch for several years, weaning some eight thousand calves each year. Cattle from the Quitaque were shipped to the ranch in the Black Hills of South Dakota.[7] During the years 1901 and 1902, Felix Franklin, a commission buyer in Amarillo, sent Cresswell and Day some thirty-five hundred head of cattle by rail to the ranch at Grand River.[8]

AGREEMENT

THIS AGREEMENT, made and entered into this 23rd day of February, A.D. 1901, by and between the Cresswell Cattle Company, a corporation, the party of the first part, and Harris, Franklin and Company, Incorporated, a corporation, party of the second part, ...

WITNESSETH:

That Whereas, the said party of the first part is now the owner of those two certain brands of cattle known as the Turkey Track [$\diagup\!\!\curlywedge\!\!\diagdown$] and the E Six [E 6] brands, said cattle ranging on what is known as the Turkey Track range in the state of South Dakota, together with said brands and the range rights to said range, and is desirous of selling and disposing of the same, This Agreement. . . .

WITNESSETH:

... And the said party of the second part covenants and agrees to and with said party of the first part that it will pay to said party of the first part as and for the purchase price of said cattle as hereinbefore enumerated, excluding calves of the year 1901, the sum of Thirty One and 50/100 ($31.50) Dollars per head. ...

It is further understood and agreed by and between the said parties that the said party of the first part hereby sells, assigns and transfers all its rights of possession to the range occupied by it and its said cattle. ...

THE CRESSWELL CATTLE COMPANY

By: Attest:

/s/ H. W. Cresswell /s/ A. J. Day

According to the terms of the contract, delivery of the cattle would continue through October 1902 and payments would be made as the cattle were received. Only the breeding stock were to be run through the chute, counted, and branded. The marketable steers would not be branded but tallied as they were loaded on stock cars waybilled for the Chicago stockyards.

The years from 1895 to 1902 proved to be very productive and profitable for the Cresswell Cattle Company. Cresswell spent much of his time in Chicago buying and marketing, and Day managed the operation in South Dakota and the Quitaque in Texas. According to data compiled by a nephew of Henry Cresswell, in 1902 Cresswell Cattle Company bought the C A Bar (C A) outfit from a Mr. Casey of Missouri. This was the Peñasco Land and Cattle Company with headquarters on the Lower Peñasco River in southern Lincoln County, New Mexico. It claimed forty miles of running water and C A Bar cattle mingled with Diamond A, Circle Diamond, and Flying H cattle that shared the open range from the Hondo River south to the Seven River country. C A Bar steers were trailed cross-country to Clayton, New Mexico, or Amarillo, Texas, where they were loaded aboard stock cars on the Fort Worth & Denver and shipped to the ranch in South Dakota. There they were turned out to fatten on grass before Day delivered them to the railroad stockyards in Evarts, on the east bank of the Missouri in South Dakota, as per the sale contract with Harris, Franklin and Company.

BURTON "CAP" MOSSMAN

Granville Graybeal, like Henry Cresswell, had operated a dairy in the Pueblo area and very likely financed his operation through the First National Bank of Pueblo. Graybeal and two partners sought financing from the Thatchers for a proposed cattle operation north of Phoenix in the Territory of Arizona. This was not an unusual request. Apparently John and Mahlon were venturesome with loan funds, financing the development of the frontier, as the First National's loan portfolio included mining and agricultural accounts in many distant areas.

The loan to Graybeal and his associates was approved, with signed notes and livestock as collateral. The trio bought cattle in New Mexico, branded them Bar Double O (O O) and trailed the herd to a ranch they had leased between the Verde and the Agua Fria Rivers in the Bloody Basin section of the Tonto Basin.[1]

In 1893, when Graybeal and his partners adventured into the business of raising of cattle in the Territory of Arizona, that country was indeed a frontier, in every sense of the word. Topographically it varied from the Sonoran Desert in the south to the forested Mogollon Rim and Colorado Plateau in the north. The inhabitants were just as varied: Mexicans, Indians, soldiers, some intrepid Anglo settlers, stockmen, prospectors, and desperados.

In the heart of this primitive territory lay the Tonto Basin. North of the basin, flanking the right-of-way of the Santa Fe Railroad, the Aztec Land and Cattle Company, the Hashknife, had established an extensive cattle operation. The two-million-acre ranch had been carved out of the heart of the territorial lands by manipulating and jawboning in the board rooms of the Atlantic & Pacific, the Santa Fe Railroad, and the Seligman brothers' bank in New York. The Hashknife, by virtue of its stockholders also being directors of the two railroads, controlled government grants to the railroads. The company also claimed range rights on hundreds of thousands of adjoining acres through the forces of money and hired gunmen.

On the south side of the Tonto Basin, the deadly feud between the Grahams and the Tewksburys over rustling, grass, and water finally ran short of participants in 1892 when Ed Tewksbury killed Tom, the last of the Graham brothers. Accounts differ, ranging from fifteen to thirty-one participants killed and an untold number wounded or disappeared.[2] Early settlers in the Tonto Basin north of Phoenix had named that area Pleasant Valley for the gentle nature of the grassy area shadowed by stately pines and watered by tributaries of the Verde and Agua Fria Rivers. The carnage between the Grahams and Tewksburys tainted the reputation of the area and the name was changed to the more fitting Bloody Basin.

Graybeal and his partners located some twelve thousand head of cattle on their ranch along the Verde River. Rustlers welcomed the ex-dairyman, his partners, and their livestock. Night-riders moved in to prowl among the Bar Double O cattle like a pack of coyotes circling the hen house. A brand on a horse or cow in that lawless area only protected ownership for those capable of out-toughing and out-riding the rustlers. Apparently the owners of the Bar Double O did not measure up. And it's quite likely that the Bar Double O cowboys were in cahoots with the rustlers. Letters from Graybeal to his bankers in Pueblo repeatedly told of finding the tracks of shod horses over those of cattle, showing where the bank's security had been driven off the ranch, and of cattle counts that were coming up short.

It was the decision of the board of directors of the First National Bank of Pueblo that the loan should be called. Graybeal and his partners were not only unable to stop the rustling but lacked the ability to drive the cattle across the Sonoran Desert to Tucson. The borrowers were not able to pay off their obligation to the bank, and as they were unable to deliver the cattle, they simply yielded to the terms of the mortgage, relinquished their rights to the mortgaged property, and dropped the keys to their Bloody Basin property in the Thatchers' laps.

John and Mahlon realized that in order to effect debt repayment through foreclosure and sale of the security they would need to send a representative to Arizona, a man who could supervise the gathering of the cattle and trail them to a railroad siding for shipment. The matter was taken up with Frank, who in his fiftieth year was a little long in the tooth for such an undertaking, and was too busy with his many and scattered responsibilities overseeing Circle Diamond and Diamond A ranches. He did undertake the task of screening and hiring a qualified man for the job.

Burton C. Mossman came west from his birthplace in Illinois by way of Minnesota. He landed in New Mexico in 1882, and, like most young men who migrated to the cattle country in those days, hired out as a cowboy. By 1889, he had advanced from being a thirty-dollar-a-month cowboy to the position of foreman of the Monticello Land and Cattle Company's Bar A Bar ranch. His employers were Langford and Carpenter, residents of Pueblo and undoubtedly social and business acquaintances of the Thatcher families. In 1893, Langford and Carpenter decided to liquidate their ranch holdings in New Mexico. Mossman was instructed to gather the cattle, trail them to the Santa Fe shipping pens at Engle, New Mexico, and waybill them to market.[3]

It was probably through his association in Pueblo with the Langford and Carpenter families that Mossman came to the attention of John, Mahlon, and Frank. Shortly after the close-out of the Bar A Bar, Mossman received a letter from Frank requesting that he come to Bloom's office in the bank at Trinidad. Bloom wanted to discuss a job offer.

During the interview, Bloom explained to Mossman the problem the bank was having in Arizona. Graybeal and his partners were unable to stop the movement of Bar Double O cattle into Mexico by rustlers. The cattle were their loan collateral. Would Mossman be interested in going to Arizona, gathering the cattle, delivering them to a shipping point, loading them, and waybilling them to the Circle Diamond Ranch at Thatcher, Colorado?[4]

Mossman took the job and arrived at the ranch in the fall of 1893. Hayden Justice, foreman of the ranch for Graybeal, agreed to stay on and help with the roundup. The parched desert that lay to the south between the Bloody Basin and the Southern Pacific Railroad line precluded moving the cattle in that direction. Mossman decided the only viable option was to find a route north to meet the Santa Fe line at Flagstaff. The cowboys strung a four-strand barbed wire fence around a large open valley near the headquarters. Cattle, as they were gathered, were held in this trap until there were enough to constitute a manageable herd on the long drive. Then the gate to the holding pasture would be thrown back and Hayden and the boys would push the cattle out and up the trail. They drove the herd north along the Verde Valley, through the rocky and forested Coconino and Kaibab mountains, and across a two-day waterless stretch. It was some seventy miles from the ranch to the Santa Fe's stock pens. The cattle were tallied, loaded into railroad cars, and sent on their way. The door of the stock car was shut behind the last load of cattle in the fall of 1897. The First National Bank of Pueblo not only recouped its money on the Graybeal loan but realized a thirty-eight-thousand-dollar profit from the sale of the collateral.[5]

In February of 1898, Mossman was hired as superintendent of the Aztec Land and Cattle Company. Historians, from Frazier Hunt writing his biography *Cap Mossman: Last of the Great Cowmen* (1951), to Robert Carlock's *The Hashknife, The Early Days of The Aztec Land and Cattle Company, Ltd.* (1994), differ somewhat on Mossman's duties and accomplishments during the three years he worked for the Hashknife. Due to continous, heavy losses of their cattle and horses to rustlers, the company liquidated their cattle in 1900. Burton Mossman, Ranch Superintendent and Agent severed, or was severed from, depending on the historian, his connections with the Hashknife at the end of that year.

In 1901, Mossman accepted a job as organizer and director of the Arizona Rangers, modeled after the Texas law enforcement group. His salary was one hundred dollars a month.[6] The Rangers had their work cut out for them, as the lawless element had for years been running roughshod over the law-abiding citizens of the territory. The force of fourteen lawmen drawing salaries from the Arizona Rangers saw to it that a number of hard cases died with their boots on, or occupied cells in the penitentiary at Solomonville.

Captain Mossman's most notable achievement came when he rode into Old Mexico after the notorious, cold-blooded Augustín Chacon. Mossman, with the help of Chacon's cohorts, Burt Alvord and Billy Stiles, got the drop on Chacon and covertly slipped him back across the border where the fugitive was hanged for his misdeeds.

Mossman captained the rangers until President Theodore Roosevelt selected Major Brodie, one of the Rough Riders, to be the

Burton "Cap" Mossman.

new territorial governor. Brodie wanted Tom Rynning, who had fought with him up the slope of Kettle Hill in Cuba, to captain the Arizona Rangers. Brodie accepted Mossman's resignation effective August 31, 1902.[7] Mossman gained a reputation as a tough, resolute lawman, and left with the moniker "Cap," which stayed with him for the rest of his life.

145

RANCH ON THE CHEYENNE RIVER RESERVATION

Shortly after the turn of the century, the Indian Service of the Department of the Interior began accepting lease applications from non-Indian cattlemen on the Lakota Sioux's Cheyenne River Reservation in South Dakota.[1] The allotted land lay east of Perkins and Eade counties, north of the Big Cheyenne River, west of the Missouri, and south of Standing Rock Sioux Reservation. The few Native American cattlemen running cattle and horses on their reservation were soon shut out when non-Indian cattle companies entered into lease contracts with the government. Only the Strip, as it was called, was fenced off and reserved for pasturage for the Sioux. It was 3.25 miles wide by 43 miles east to west, 139.75 sections, or 89,440 acres separating the Cheyenne River and Standing Rock Reservations.

The Hansford Land and Cattle Company, a Scottish investment group, ran cattle under the Turkey Track brand in Hutchinson and Hansford counties in Texas and in Lincoln County in New Mexico. The company was organized by and under the financial management of James Coburn, Scottish-born American and Kansas City banker.

In 1903, there were some ten thousand head of Turkey Track cattle on Hansford property in Lincoln County. Their range stretched from below the headquarters at Lakewood north some forty miles along the east side of the Pecos River. Cap Mossman was introduced to James Coburn. Coburn hired Mossman as foreman of the Turkey Track operation in New Mexico.[2]

As a guest of the Chicago, Milwaukee and St. Paul Railroad on a tour promoting settlement, Coburn traveled to the railhead at Evarts in central South Dakota.[3] The sight of waving stands of

The remuda in a rope corral on the Turkey Track Ranch in South Dakota, circa 1901.

grasses on the hills of the Cheyenne River Indian Reservation west of the Missouri, coupled with the information that bids for grazing leases on the reservation were to be accepted from interested cattlemen, caught Coburn's attention.

Murdo Mackenzie, as manager of the Scottish-owned Matador Land and Cattle Company, Ltd., moved to lease the entire east half of the Cheyenne River Reservation for the Matador's Drag V (⌐U⌐) cattle. Their lease, of about a million and a half acres, lay north of the Cheyenne River to its junction with the Missouri River, west of the Missouri, and north to the division line between the Cheyenne River and the Standing Rock Reservations.[4]

Coburn, faced with a protracted drought and an overstocked range on the Turkey Track ranch in southern New Mexico, saw a solution to the problem in the vast expanse of lush prairie he had seen in South Dakota. He met with Matador management and arranged to take over a portion of their lease. He sent Cap Mossman to meet with Mackenzie, look over the range, and agree as to the boundaries between the company leases.[5]

Under the agreement between Mackenzie and Mossman, Matador had the northeast half of the reservation and the Hansford Land and Cattle Company the 488,000 acres in the southeast half. The leases were long-term and initially cost three cents per acre

Cattle in dipping pens at Evarts, South Dakota, circa 1904. Dipping against scabies.

per year, the money going to the Bureau of Indian Affairs, an agency of the Department of the Interior. It was agreed between Matador and Hansford to let a contract for a four-wire fence running some forty miles east and west to separate the Turkey Track and Drag V cattle.[6]

In 1904, the Matador Land and Cattle Company moved their first bunch of cattle from their ranch in Texas to the lease on the Cheyenne River Reservation. The first bunch of Turkey Track cattle were trailed from Hansford Land and Cattle Company's vast spread south of Artesia, New Mexico to Amarillo, Texas, in the spring of 1904. From there the cattle were freighted to Evarts. Ranch hands rode in the caboose to accompany the shipments. It was their task to walk alongside the train when it stopped for fuel or water, jabbing with sharpened poles between the slats of the stock cars, prodding down cattle to their feet. It was also their job to pry open the stubborn sliding doors of the cars and wrestle with the heavy bull boards to unload and load the cattle at certain stops, as it was required by law that transient cattle be fed and watered every thirty-six hours.

At Evarts, the cattle were unloaded into stock pens. Those that needed to be doctored were driven from a crowding pen that funneled into a narrow pole-sided chute with a head catch and squeeze section at the far end. Then, because of an outbreak of scabies, all the cattle were hazed into a chute leading to a long, deep, water-filled concrete vat. The dip was charged with either sulfur and lime or a heated solution of Black Leaf 40, a tobacco derivative. Cowboys stood on each side of the vat goading the swimming cattle to the far end. They used long poles with a large, lazy-S-shaped iron rod attached at right angles to the bottom of the pole. The down-curve of the S was used to push the cattle down into the medicated solution, being sure that the animal, head and all, was submerged. The up-curve of the S was employed if an animal began to show signs of drowning. It would be hooked under the neck to pull the head up so the animal could breathe. The panicked cattle crowded and trampled over one another. This was an ordeal for stock weakened from long, crowded confinement. At times, the crews stopped to drag the dead animals out of the vat.

After they had been "worked," the Turkey Track cattle were herded into a funnel-shaped pen that led to the river. They were loaded onto barges with high railings and floated across the Missouri River to the Strip. If the river was low, cattle might be crowded in and made to swim across, or a pontoon bridge might be the means of crossing. By the terms of the agreement between Mackenzie and Mossman, the Turkey Track cattle were then trailed south across the Drag V country to the Hansford lease.

After their meeting in Coburn's Kansas City office in 1903, during which they discussed the terms of Mossman's employment and his duties and responsibilities on the Turkey Track, the two men had retired to Coburn's home for dinner. Cap met and was attracted to Coburn's daughter, Grace. In the fall of 1905, Burton Mossman asked Coburn for Grace's hand in marriage. The young couple were married that December.[7]

Coburn turned the Hansford Land and Cattle Company's lease on Cheyenne River Reservation over to his new son-in-law. It was agreed that Mossman would pasture Hansford's Turkey Track cattle on a cost-per-head basis. Pastured cattle, a cow and a calf, or a steer—long-yearling or older—were bringing five dollars a head for a two-year grazing period, or three dollars for one year.[8]

Ike Blasingame, a cowboy for the Matador, in his book, *Dakota Cowboy,* described Cap Mossman from their first meeting in Evarts.

> The first glimpse I ever had of Cap Mossman, manager of the Turkey Tracks, was in Joe Green's hotel in Evarts in the spring of 1904. He was a young man, not too tall, and wore a black moustache. His gray hat was set at a jaunty angle on his head, and from under its broad brim his blue eyes had a way of looking sharply at each and every man he met. He was a colorful character. . . .
> He was standing at the hotel bar with a drink before him, lighting his cigar with a hundred-dollar bill! He folded the greenback lengthwise several times, then, reaching over to the cigar lighter which always stood on the bar for customers' use, Cap dipped the end of the bill into the flame. When it blazed up good, he put it to his cigar. . . . He pinched out the fire, pocketed the bill, puffed his cigar, and took up his drink as if lighting cigars with a big greenback was an everyday occurrence with him.[9]

In 1907, Cap, with income from pasturing Hansford Land and Cattle Company's cattle, increased his lease holding on the reservation by taking over the seven-hundred-thousand-acre lease held by the White River Cattle Company, known by its Sword and Dagger brand.[10] Along with several smaller leases he had acquired, this put Mossman's holdings at over 1.25 million acres. His leases were costing him fifty-three thousand dollars a year.[11] He negotiated with Colonel Green, an acquaintance with far-flung investments, for financing. The disastrous winter of 1906–07, coupled with a downturn of the national economy in 1907, caused Green to discontinue his financial backing. This left Mossman with a large debt and a bear cattle market. Looking for help he contacted Mahlon Thatcher, who was in Bon Air, Georgia, and arranged to meet with him and Frank Bloom in Kansas City. Bloom then accompanied Mossman to South Dakota where he looked over the operation.[12] He carried a favorable report back to John and Mahlon.

Possibly because of the increased interest and activity of the Thatchers in acquiring land, the Bloom Cattle Company corporate name was changed in 1907 to the Bloom Land and Cattle Company. In June of the same year, the Diamond A Cattle

Company was incorporated. The object of the Diamond A Cattle Company was "engaging in a general land and livestock business." The capital stock of the company was $250,000, divided into twenty-five hundred shares. The operations of the company were to be carried on "in the County of Pueblo, State of Colorado and also in the State of South Dakota." The business office was located in Pueblo. John A. Thatcher was president; Frank Bloom, vice president; A. S. Booth, secretary; Mahlon D. Thatcher, treasurer, and B. C. Mossman, general manager.[13]

Mossman's salary was set at a hundred dollars a month and he was instructed to "arrange for the leasing of such lands as was required in South Dakota and the purchase of sufficient cattle with which to stock same, along with the necessary horses and equipment to properly take care of the Diamond A Cattle Company's interests."[14]

The South Dakota ranch, under Cap Mossman as general manager and John Bloom, Frank's nephew, as resident manager, grazed steers that Frank sent from other company ranches and cattle he purchased. Commission men were engaged to purchase steers to stock the Cheyenne River Reservation lease. There were THS-branded longhorns from the empire of Don Luis Terrazas in El Estado Chihuahua, Mexico,[15] and cattle belonging to Mossman's acquaintance, Col. Green.

Fifty to eighty cowboys rode herd on fifty thousand steers that were put in as yearlings, run for three years, then sent to market in Chicago. After a few years, the company changed the stocking of the South Dakota ranch by supplying their own livestock, which included raised and purchased steers, a herd of breeding cattle, and a large band of brood mares and stallions.

A profit and loss statement dated December 31, 1912, for the Diamond A's operation in South Dakota showed a net profit of $118,341.43. On January 2, 1914, the Bureau of Indian Affairs notified the company that part of Mossman's lease in the southwestern corner of the reservation, including the Sword and Dagger country, would not be renewed. That part of the Sioux tribes' reservation was to be opened to non-Indian settlers. This left something like 750,000 acres for Diamond A cattle.

At a meeting held in Pueblo on January 17, 1914, the directors voted to shore up by leasing land owned by W. E. Harris in Custer County, Montana, for a period of five years starting May 1, 1914 at

Bloom Ranch, Wagon Mound, New Mexico. Frank Bloom visits the Diamond A Wagon Mound Ranch in 1926.

an annual rental of five thousand dollars. This lease was along the Powder River some sixty miles below where it flowed into the Yellowstone. Mossman moved some fifteen thousand cattle off the South Dakota country to the Montana lease.

Cap's wife, Grace, passed away a few days after the birth of Mary, their second child. The Hansford Land and Cattle Company had decided in 1912 to sell their Turkey Track ranch in New Mexico. With his father-in-law's help, Mossman arranged to buy the deeded headquarters area and take over the five-hundred-thousand-acre grazing lease.[16]

In 1917, the Bloom Land and Cattle Company began acquiring deeded land west of the village of Wagon Mound, New Mexico. The Bloom Ranch, as it was called, was part of the Mora Land Grant; a colonization grant issued in 1835 by the Republic of México responding to a petition by seventy-six Mexican citizens, residents of the Mora Valley in the Sangre de Cristo Mountains. Including the *ejido*, common lands, the grant covered 827,621 acres. The grant was patented by the U. S. government in 1860. The company put together a ranch unit of 94,386 acres. In the days of the Santa Fe Trail, wagons headed by the way of Fort Union left deep wheel ruts across the Bloom Ranch.

In 1919, the directors voted to increase Mossman's salary to five thousand dollars per annum and renew the Montana lease for

Bloom Ranch Sheep Camp, Wagon Mound, New Mexico, circa 1926.

two years with an option for three more years at an annual rental of three thousand dollars. Over Mossman's objections, the company ventured into the sheep business on their Montana lease. Over half of the flock was lost during the winter of 1916. But favorable prices for wool after World War I recouped the investment, with interest.

In 1923, the company decided to make combination cattle, sheep, and horse ranches out of the Bloom Ranch near Wagon Mound and Circle Diamond west of Roswell. Within a few years the sheep inventory on the two ranches exceeded fifty thousand head.[17]

Overextension of credit and unfavorable years for cattle production put Cap Mossman in a financial bind on his personal ranching operations. In 1925, the Diamond A Cattle Company took over the Turkey Track Ranch east of Artesia, New Mexico along with Cap's sixteen thousand head of cattle, his personal leases in Lincoln County, and fifteen hundred head of breeding cattle in South Dakota.[18]

At a stockholders' special meeting held in Pueblo, Colorado on December 31, 1938, it was resolved:

> That The Diamond A Cattle Company be completely liquidated . . . by the assignment, transfer and conveyance as of this date of all of its assets to the Bloom Land and Cattle Company. . .

> Deeds prepared and executed today to convey to the Bloom Land and Cattle Company title to all tracts of real estate owned by the Diamond A Cattle Company in the states of New Mexico and South

Poker and three-card monte were the most popular card games of chance among cowboys. Circle Diamond Ranch near Malta, Montana. Courtesy Montana Historical Society.

Dakota, and bills of sale prepared and executed to convey to the Bloom
Land and Cattle Company ownership to personal property of wells,
pumping equipment, tools, wagons, automobiles, and livestock consisting of horses, mules, cattle, and sheep owned by the Diamond A
Cattle Company in the States of New Mexico and South Dakota and all
livestock consisting of cattle owned by the Diamond A Cattle Company in the State of Montana.[19]

THE HATCHET CATTLE COMPANY

The expanse of public domain on the Great Plains was no longer a sovereignty for the Goliaths of the livestock industry. David, with a plow in place of a sling, followed the railroads into the midcontinent and the family farm began to take root. Stockmen with foresight, like those with the Bloom Land and Cattle Company and the Matador Land and Cattle Company, Ltd., could see the end of an era. They were capitalizing on a frontier that was still beyond David's reach and establishing solid bases for their ranching operations by buying and leasing land, putting it under fence, and drilling wells for water.

The absentee investor was learning the hard way that there was more to ranching than turning livestock out onto the range and sitting back to reap the profits. The cyclical market for cattle and sheep affected income and dividends. Factors that influenced the market, such as supply and demand or the projected corn crop in Iowa, were studied by astute stockmen who strove to buy their cattle or sheep on a bear market and sell on a bull market. Sound advise for stockmen was to sell your production and buy your replacements on the same market cycle.

Reports of death losses after the extremes of winter were catastrophic proof of the adage that you can't starve a profit out of a cow. Venture capitalists from the boardrooms back East and abroad who had investments in the livestock industry on the Great Plains learned during the decade of the 1880s that roses do indeed have thorns. Graziers were becoming painfully aware that the ecology of the prairie was fragile, and there were limits to the number of livestock their ranges could support and still remain viable. They learned that some country was best suited for summer grazing, whereas areas that provided natural protection from

Haying on Bloom Ranch, Wagon Mound, New Mexico, circa 1930.

the elements were better used for winter pastures. Knowledgeable stockmen looked for ranches consisting of a balance between the two. Supplemental feeding of livestock with hay during stormy winter months and during the calving or lambing season was becoming an essential part of husbandry, and an accepted part of the overhead. Successful stockmen saw the necessity of improving the quality, uniformity, and predictability of their herd by using pure-blooded sires and keeping the top end of their daughters for replacements. Charles Goodnight, Frank Bloom, Henry Cresswell, and Tony Day were industry leaders in progressive animal husbandry. Going into the twentieth century, stockmen found that herd improvement and sound, progressive ranch management were necessary for survival.

Even the hardscrabble but happy-go-lucky life of the cowboy was changing. No longer did he spend his days sitting with his forked end astraddle a horse on a trail drive to the bright lights of Dodge or just prowl the open range on ol' Socks. Now he might be called upon to ride on the bouncing, iron seat of a mowing machine or a rake in a hay field, climb a windmill tower to oil the gears, or wear out his gloves digging post holes and stretching that curse of the open range, barbed wire. During the winter months he could no longer hole up in a line camp, only needing to keep a supply of wood at hand for his fire. It fell to his lot now to put on his sheepskin coat and four-buckle overboots, harness and hook up the team, load the wagon with hay, buck the crusted snow and drifts to carry feed, and chop holes in the ice covering the watering holes for cattle.

Public lands that were still available on the perimeters of the frontier were being tied up with long-term leases by operators who had adjoining deeded land. The Thatchers and their associates were adapting to the dictates of the time. In southeastern Colorado, the Bloom Land and Cattle Company grazed thousands of steers as well as some two thousand head of horses.[1] Circle Diamond cowboys might water their mounts in the Purgatory River where it coursed through the prairie northeast of Trinidad, or roll their beds out under a stately blue spruce a hundred miles away in the evening shadow of Wet Mountain in the Sangre de Cristos. The Thatchers and Frank Bloom, operating the stock-raising enterprise from the company's ranch headquarters at Thatcher, utilized a vast range of public domain that was checkerboarded and held together with deeded acreages. John, Mahlon, and A. S. Booth, their bookkeeper, administered the company's business in the First National Bank of Pueblo. Frank and his wife Sallie lived in Trinidad, on those rare occasions when he was not at, or on the road to, one or another of the company's far-flung ranches.

✳

Carl Stanley's Two Quarter Circles Ranch () with its headquarters on Greenhorn River claimed, according to his letterhead, "a range from the Arkansas River to the Apishapa Creek" in Pueblo County, Colorado.[2] He obtained financing from the First National Bank of Pueblo. In 1884, Stanley bought the Hatchet Ranch in the territory of New Mexico from the Rechter brothers. The ranch was almost entirely public domain. It was in the extreme southwest corner, just a few miles north of the Mexican border. The natural barriers to the roaming of cattle were the rugged Peloncillo Mountains on the west and the Cedar Mountain Range on the east. The headquarters to this 1.6 million acres in the heart of the Chihuahuan Desert was in the Hatchet Gap between the Pyramid Mountains and Sierra Hatchita.

ARTICLES OF INCORPORATION

KNOW ALL MEN BY THESE PRESENT that we, Carl Stanley, Mahlon T. Everhart and A.S. Booth, residents of the State of Colorado, have associated ourselves together as a company under the name and style of The Hatchet Cattle Company, for the purpose of becoming a body corporate and politic under and by virtue of the laws of the State of Colorado and in accordance with the provisions of the laws. . . .
    2. Our said company is formed and incorporated for the purpose of raising, buying and selling cattle, for the purpose of acquiring pasturage and grazing lands for the same, ...
    3. The capital stock of our said company is One Hundred Thousand Dollars, ...
    6. The operations of our said company will be carried on in the County of Grant, Territory of New Mexico and in the County of Pueblo, State of Colorado, and the principal place and business office of said company shall be located in the City of Pueblo, County of Pueblo and State of Colorado. . . .

In Witness Whereof we have hereunto set our hands and seals this 16th day of October, A.D. 1902.[3]

Mary C. Thatcher, John and Mahlon's sister and Frank's sister-in-law, married Marshall H. Everhart of Pennsylvania. A son, Mahlon T. Everhart, was born March 24, 1873. He came to Colorado as a youth to learn the ranching business from his uncles. Frank took young Mahlon under his wing and was the young man's mentor. In 1898, Mahlon T. partnered with his uncles, John A. and Mahlon D., in the formation of the Colorado Arizona Sheep Company. Young Everhart put up a third and his uncles two-thirds of the capital necessary to buy a band of sheep in Folsom, New Mexico. Mahlon T., with a crew of men, railed and trailed the sheep to a leased ranch near Springerville, Arizona. Mahlon looked after this investment until sometime in 1902, when it was liquidated and the proceeds invested in the Hatchet Cattle Company.

Albert S. Booth was born May 20, 1866, in Illinois. He went to work for the First National Bank of Pueblo as secretary to Mahlon D. Thatcher, then president of the bank. On March 15, 1902, Booth was promoted to assistant cashier. Booth was probably not an investor in the Hatchet Cattle Company himself, but as a trusted director of the Hatchet Cattle Company and an employee of the Thatchers' bank, he acted as a liaison between the two organizations.

In 1905, the Thatchers and Mahlon Everhart bought out Carl Stanley's extensive real estate and livestock holdings in Colorado and New Mexico and his interest in the Hatchet Cattle Company. They named the purchased property along the Muddy Creek in Pueblo County the Hatchet Ranch, and that on the Greenhorn River they called the Red Top. With Stanley out of the picture, The Hatchet Cattle Company was a three-way partnership between Mahlon and his two uncles. Mahlon D. Thatcher was president. Their brand was the Hatchet ($\triangle$).

In 1902, Albert Fall, an attorney in New Mexico, began acquiring acreage along the Three Rivers area in Lincoln County. In 1907, he acquired Pat Coghlan's 103,000-acre Tres Ritos Ranch, lying between the Mescalero Apache Reservation and the El Paso and Northeastern Railroad right-of-way.

In 1909, Fall was in court in El Paso defending Wayne Brazil, who was accused of killing Pat Garrett. Mahlon Everhart was also in El Paso on behalf of his foreman on the Hatchet Ranch, who was accused of illegally crossing cattle back and forth across the Mexico–United States border. Mahlon met Carolyn Ann Fall, who was in El Paso visiting her father. Later that year, Mahlon proposed to "Carrie" on the steps of governor's mansion in Santa Fe. They were married in December of 1909. The following year, Mahlon went into partnership with Albert Fall, and the Tres Ritos Land and Cattle Company was incorporated. Fall was president and Mahlon was vice president. They branded R on the left shoulder, R on the left rib, and R on the left hip. The ranch was called the Three Rs.[4]

In the early 1880s, the Harris-Brownfield Bar W cattle ranged along the Tularosa Basin from south of White Oaks in the Carrizozo Mountains of Lincoln County almost to El Paso. In 1912, the Hatchet Cattle Company purchased the Bar W's deeded properties, with watering rights. This gave the Thatchers and Mahlon Everhart control of deeded and range rights amounting to some nine hundred thousand acres west of the El Paso & Northeastern Railroad's line running along the Tularosa Basin between Carrizozo and El Paso.[5] This was a steer operation using the Stroke Box brand (/$\square$), which was registered for the Hatchet Cattle

Company in Colorado and New Mexico. Mahlon Everhart was the president.

Fall added to his Three Rivers property by acquiring Monroe Harper's ranch in 1915. He also bought half of Pete Crawford's ranch and all his water rights, which Fall sold to the El Paso & Northeastern Railroad for seventy-five thousand dollars. Fall's Three Rivers Ranch and Fall and Everhart's partnership property, the Tres Ritos Land and Cattle Company, neighbored the Hatchet Cattle Company, separated by the railroad's right-of-way. United by family ties, the Hatchet Cattle Company, the Three Rivers Ranch, and the Tres Ritos Land and Cattle Company controlled a range that extended over a hundred miles along the Tularosa Basin between the Organ and San Andreas chain of mountains and the Sacramento Mountains, encompassing something over a million acres.

In 1921, Fall was appointed Secretary of the Interior by President Harding. While a member of the cabinet, Fall sold an interest in the Three Rivers Ranch to Harry Sinclair of Sinclair Oil Company and borrowed money to finance the addition of land and water rights to the ranch. The deal with Sinclair, and a top-heavy loan from Ed Doheny of the Pan American Oil Company, ultimately led to Fall being implicated in the Senate's investigation of the Department of Interior's impropriety in the handling of naval oil reserves. The prosecutor for the government argued that Sinclair's financial involvement and Doheny's loan were actually bribes soliciting favorable drilling rights for Sinclair Oil Company and Pan American Oil Company from Secretary of the Interior Albert B. Fall, in the Teapot Dome and Elk Hill government-held oil reserves. Fall was convicted and sent to prison in 1931.

Doheny's heirs foreclosed their mortgage on Fall's Three Rivers Ranch. The Hatchet Cattle Company reacted at a called meeting:

DIRECTOR'S SPECIAL MEETING
THE HATCHET CATTLE COMPANY
Pueblo, Colorado, August 6, 1935

A special meeting of the Board of Directors of The Hatchet Cattle Company was held at the office of the company in the Thatcher

Building, Pueblo, Colorado, at eleven o'clock A.M. on this Tuesday, August 6, 1935, all members of the Board being present.

M. T. Everhart and L. T. Rule acted respectively as Chairman and Secretary of the meeting.

The Chairman explained that the meeting had been called for the purpose of considering sale to Tres Ritos Ranch Company of the Hatchet Cattle Company's Three Rivers Ranch Properties, as contemplated in agreement dated August 1, 1935 between this company and Mahlon T. Everhart as Sellors and Tres Ritos Ranch Company as Buyer.

After general discussion of the proposed sale and the further explanation by the Chairman that it has been understood with the Petroleum Securities Company interests that in event of the sale of their Three Rivers holdings the ranch properties used by the Hatchet Cattle Company in its Three Rivers operations would be included for consideration of $30,000, the following resolution was moved, duly seconded, and upon being put to vote unanimously adopted:

WHEREAS, there has been presented to and read at this meeting, a form of a proposed agreement dated August 1, 1935, between the Hatchet Cattle Company, a Colorado corporation, and Mahlon T. Everhart, parties of the first part, and Tres Ritos Ranch Company, a New Mexico corporation, party of the second part, in and by the terms of which parties of the first part agree to sell, assign, transfer and convey unto party of the second part, all of their cattle ranch properties.[6]

Mahlon returned to Colorado where he supervised the Red Top and Hatchet Ranches south of Pueblo and the Three R Ranch on Wet Mountain near Beulah.

FEW EQUALS AND
NO SUPERIORS

Through the long lane of passing years I see Mr. Cresswell sitting at his desk in our office, writing up his little book, scrupulously accurate, not an item escaping him. At the end of the season Tony came to Chicago, settled accounts, probably took a short trip, then back to work. That Cresswell & Day firm was a great combination. It was teamwork.[1]

After closing out the South Dakota operation in 1902, Cresswell and Day moved cattle from the Lazy F in the Texas Panhandle and the C A Bar in southern New Mexico to a leased place near Billings, Montana, where they remained for winter. Early in the spring of 1903, they threw their cattle and horses on the trail to Canada. They had negotiated with the Canadian government to lease a spread thirty-six miles wide by forty-two miles long—967,680 acres or forty-two townships—southeast of Swift Current, at two cents an acre.[2]

Cresswell Cattle Company cowboys pushed cattle along a trace that circumvented settlers' barbed-wire fences and forded icy rivers for some two hundred miles across Montana, then trekked north over the Canadian prairie for some three hundred miles through the sparsely-inhabited southwest corner of Saskatchewan to their new location on Swift Current Creek.

The partners began to freight cattle from their operations in Texas and New Mexico to Canada. Cattle from the Lazy F Ranch were loaded in stock cars at Groom, Texas, and hauled via the Rock Island, Northern Pacific, Chicago-Milwaukee and Canadian Pacific railroads, by way of Lincoln and Omaha, Nebraska; Council Bluffs, Iowa; and Aberdeen, South Dakota to the stock pens at Swift Current, Saskatchewan.[3] From there the cattle were trailed south to the ranch. From the C A Bar, cattle were thrown on the trail and herded some two hundred fifty miles to Amarillo, Texas, or Clayton, New Mexico, then railed to Canada.

Cresswell bought twenty thousand head of top-of-the-line breeding Hereford cattle from Charlie Goodnight, who supervised the loading of the cattle into stock cars at Groom, Texas, and waybilled them to the Cresswell Cattle Company at Medicine Hat in Alberta. From there they were freighted on the Canadian Pacific to Swift Current, then trailed to the ranch. Later, seven thousand cattle branded 7 D were purchased near Dalhart, Texas, and sent by rail to Swift Current.[4] Within a year the Turkey Track was running seven hundred horses and over thirty thousand head of cattle, of which some eight thousand were young, light-weight cattle that had been purchased in Manitoba.[5] A severe storm in May 1903 killed about one-fourth of those young cattle.

The January 29, 1905 *Pueblo Chieftain* reported:

BLOOD POISON CAUSES DEATH
OF FORMER PUEBLO PIONEER

Henry W. Cresswell was for Years One of Colorado's Big Cattlemen.

After suffering for several months from an injury which at first appeared to be most trivial, and twice having a portion of his limb amputated, Henry W. Cresswell, a pioneer and for years a resident of Pueblo, yesterday died of blood poisoning. . . .

Mr. Cresswell has for years been known as one of the big cattle kings of the west and at the time of his death was in charge of big cattle interests in Northwest territory, Canada, in which the Thatcher brothers of Pueblo are interested.

The *Medicine Hat News,* of Alberta, Canada carried the following in its February 2, 1905 edition:

John A. Thatcher, of Pueblo, Colorado, who is a member of the Day-Cresswell ranching firm, was at the bedside of his long-time friend and business associate Henry Cresswell.

In *Memory Cups of Panhandle Pioneers,* by Millie Jones Porter, E. H. Brainerd, who started to work on the Bar CC when he was twenty-one years old, is quoted as saying of Henry Cresswell: "No better man ever lived than Hank Cresswell."[6] J. Evetts Haley in his biography of Charles Goodnight quotes Goodnight as saying, "Among those who should be remembered for their work in bringing law to southern Colorado was . . . H. W. Cresswell."[7]

John A. Thatcher named a son Raymond Calvin Cresswell Thatcher. John Clay of John Clay and Company, a man who had dealings with every cattleman in the Southwest, spoke of Hank during his Panhandle ranch days: "Cresswell is a range cattleman with few equals and no superiors."[8]

Henry Cresswell was indeed a man "chiseled out of the old rock," to use an analogy from Frank Dobie's writings. In the lingo of men of the range, Hank Cresswell was "a man to ride the river with," a steadfast and trustworthy partner; not only a successful cattleman but a man who could make a hand with the best of them. He was like Albert K. Mitchell of the Tesquesquite Ranch in Harding County, New Mexico, who was highly respected for his ability to manage vast ranching operations as well as working in the branding corral. As a cowboy dragging calves for a branding crew, Mitchell could pick up both heels with every loop.

Cresswell met the rigors of frontier life head-on in establishing the Bar CC in Texas and the Turkey Tracks in Oklahoma, South Dakota, Montana, and Canada. According to Laura V. Hamner in her biography, titled *A Batchelor's Progress:*

> Thus ended Hank Cresswell's cattle history. Hank had pressed his brand on the Bar CC range and on the pastures of the Indian Territory, South Dakota, Montana, and Saskatchewan. The Thatchers were always his silent partners, and later Tony Day was co-owner and manager with him. . . .
>
> He had lived a full, rich life, robust and vigorous to the last. He had made money on every venture.[9]

But it is not Hank Cresswell's financial success that people remember; it is as a maker of men that he is best known. He insisted that his cowboys file on land, that they start a herd. "There's room and grass for all of us," he often said. No one knew all the people that Cresswell helped, for he never boasted. No one doubts that many of his men owe their later financial success to the aid and encouragement of their boss.

In 1903, O. H. Perry Baxter had sold his interest in the Cresswell Cattle Company to his partners. He was involved in real estate development in Pueblo County and the state of Colorado. He passed away in 1910 at the age of seventy-five.

The Cresswell Cattle Company continued under Tony Day's management. Going into the winter of 1906 they were running about twenty-five thousand head of breeding cattle and beef steers. Winter started early in October with a three-day blizzard. Cattle drifted for miles ahead of the biting winds, hunting protection. A short, balmy Indian summer was followed in November by another blizzard. By the middle of December, the snow was a foot or more deep on the level and drifts were up to the eaves of the cabins. Shortly before Christmas, the temperature sank to twenty degrees below zero. Cattle were freezing or starving to death by the thousands.

After the weather abated, Tony Day and his nephew, John Day, put a camp outfit in a buckboard and tied a saddle pony alongside and started out to look the situation over. The snow was, by now, melted, and what they saw left no doubt of the loss they might expect. Tony counted some three hundred carcasses in a ride of less than two miles across the prairie, over five hundred down along the river flats, and eighteen more that had sought shelter in a log shack where Tony and John had intended to spend the night.[10]

The remnants of the Turkey Track cattle were gathered during the roundup in the spring from as far away as Wood Mountain and the Big Muddy country, some seventy to ninety miles southeast of the ranch headquarters. Day reported to the main office in Pueblo that he estimated the loss of cattle would run close to sixty percent. A roundup and tally, after all the ranches in the area had completed their roundups, put Turkey Track losses at nearer seventy percent. The Cresswell Cattle Company sold out its Canadian holdings in the summer of 1908.

In 1916, Day sold his horse herd of twelve hundred head and moved to California. He died there in 1928 at the age of seventy-nine.

CIRCLE DIAMOND AND
TEE DOWN BAR

Whoopee ti yi yo, git along little dogies,
It's your misfortune, and none of my own.
Whoopie ti yi yo, git along little dogies,
For you know Montana will be your new home.

R. M. Allen, Harvard University alumnus and manager of the
large holdings of the Standard Cattle Company, was asked in 1893
by Harvard to assess opportunities in the range cattle industry for
college graduates. He suggested that should a graduate be interested in ranching, he should first consider hiring on as a cowboy
to learn the practical part of the business. The expense of establishing a ranch was so high and the risk of loss so great as to be
prohibitive. Besides, Allen pointed out, the only available range
left was in Montana.[1]

Frank Bloom located the Circle Diamond headquarters six
miles north of Malta, Montana, on Assiniboine Creek. From a
deeded land base of 439.97 acres of scattered parcels of land,[2] the
Bloom Cattle Company laid claim to range rights from Hinsdale
to Chinook and from the Milk River to the Canadian border,[3]
roughly 2.25 million acres.

"Way up north," as the musical refrain goes, was the last frontier on the midcontinent where grass blanketed hills and valleys.
A vacated region of public domain dotted with circles of stones
that had held down the bottoms of lodge covers and charcoal bits
from long-dead campfires lay waiting for cattlemen with the
money and muscle to claim range rights in the old way.

Frank Bloom sent the first herd of cattle to Montana from the
southern ranches in 1892. Early on, the Circle Diamond was
strictly a steer and horse breeding operation. Frank would ship
from seven to ten thousand steers each spring to John Survant at

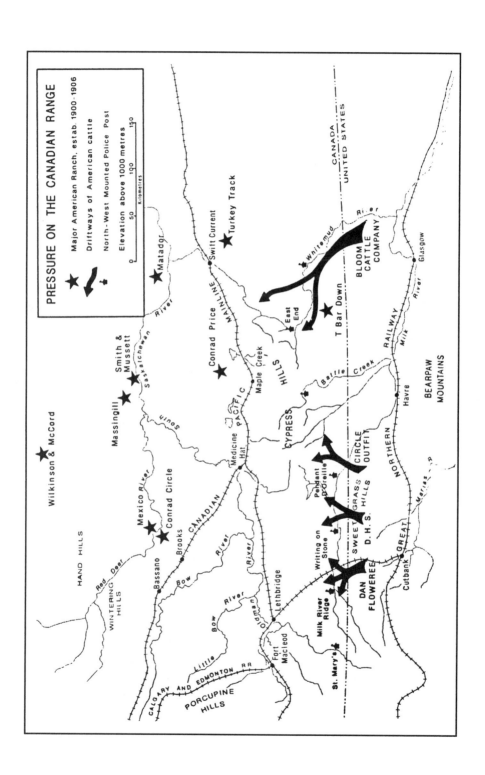

PRESSURE ON THE CANADIAN RANGE

★ Major American Ranch, estab. 1900-1906

↙ Driftways of American cattle

⌖ North-West Mounted Police Post

Elevation above 1000 metres

0   50   100   150
Kilometres

the Circle Diamond headquarters in Malta. Some of these cattle came from company breeding ranches. Bloom also bought cattle. Correspondence from Bloom to John Thatcher shows that he was receiving cattle for the Montana operation in the Wilcox, Arizona country and off ranches west of Magdalena, New Mexico. These cattle were trailed along the Magdalena Stock Driveway to the railroad loading pens in Magdalena, in its heyday the third largest livestock shipping point in the world. The cattle were railed to Orin Junction in Wyoming and then trailed some four hundred miles to the ranch at Malta.

In the fall, hopefully before the first heavy snow, Survant and his cowboys would gather and sort out marketable beeves, the three- and four-year-olds, and any remnant steers that had been missed the previous year, and trail them off to market. After the Northern Pacific Railroad laid tracks through Montana in 1903, the shipping and receiving of Circle Diamond cattle was done at Billings. From Circle Diamond headquarters to Billings was a two-hundred-plus mile trail drive with the wide and sometimes treacherous Missouri River in between. In a letter to John Thatcher, Bloom mentioned the fact that Survant had reported losing fifteen steers during a crossing of the river. In *Campfires and Cowchips*, Floyd Hardin recalls John Survant's story about crossing the Missouri:

> Oren Backues and I started out from Malta to the Missouri River to meet a herd of Circle Diamond cattle coming over from Billings. It was in June, and all the creeks were bank full. We started early in the morning, hit Big Warm Creek about noon, and found it out of its banks, from hill to hill. It was either swim or ride twenty miles to the Phillips Ranch, then across to the George Jones ranch on Little Warm, so we decided to swim.
>
> Stripping down to our birthdays suits, we tied our clothes behind our saddles, then rode out to swimming water, slid off and caught our horses' tails, and swam across behind them. When we reached the bank and came out on dry ground, the horses took one look at us, let out a startled snort and took off across the prairie, leaving us standing in the middle of nowhere with not so much as a cigarette paper for cover. There was only one thing for us to do, set out on foot for the George Jones ranch some twenty miles away.
>
> With Oren in the lead, we dodged cactus and stumbled through sagebrush all afternoon. When possible, we followed cattle trails, but these often changed directions on us, and we had to cut across the trackless prairie. It was a bright, warm day, but as the sun sank toward

the horizon, the cold evening air began to add to our sufferings. About sundown, we hit an abandoned sheep camp. Oren found an old, worn out pair of overshoes, and I robbed a scarecrow of an old pair of torn and windblown overalls. We put these on and stumbled on through the night, finally hit the Jones ranch sometime after midnight.

The bristling ranch dogs didn't like us any better than our horses, and stood us off from the buildings with their growling and barking while we stood half frozen, adding our yells to the din, until the noise finally woke up some of the ranch hands who quieted the dogs until we could get in. We were cold, tired, hungry, and rather unhappy about the whole deal, and those d.... dogs didn't help any. If we'd had guns Jones would've needed a new bunch next morning.

We spent the next day and night resting. Jones fixed us up with some old clothes, but our feet were full of cactus spines, and swollen so badly we couldn't wear shoes, so we wrapped them in gunny sacks. Jones was a sheepman, so didn't have too many horses, but staked us to a couple of crow-baits used by the herders, but no saddles, so we started for Malta bareback.

At the Phillips ranch, some fifteen miles along the road, they managed to scare up a couple of old saddles for us. We still couldn't get our bundled feet in the stirrups, but it beat riding bareback.

We got to Malta just in time to stop a search party which was ready to start out looking for our bodies. A Circle C rider picked up our horses with our saddles and clothes, brought them to Malta and reported us drowned.

I don't know what kept Oren going that night, but if I hadn't been behind watching him dodge cactus and sagebrush, I'd never have made it.[4]

Malta was a typical frontier cow town; a cluster of log buildings hunkered beside the rails of the Great Northern Pacific Railroad, north of the Milk River. "A tough little town," according to Floyd Hardin, a Circle Diamond 'puncher who worked under John Survant. Most of the business places, with their false fronts, were strung out along the road north of the railroad tracks. Flanking a boxcar, grounded beside the tracks with the word DEPOT painted over the door and MALTA on each end, were the ever-present business houses: a livery stable, feed barn, a blacksmith, the Malta House which offered rooms and board, a general store, a "greasy spoon" cafe, six saloons, several residential homes, and three "pleasure palaces."[5]

Cattle ranching was moving beyond the trail driving, cattle baron, free-range era. But the basics of cattle husbandry remained the same. No matter how tall or lush the grass, it was of no use to a bovine without adequate water somewhere in the proximity. It still took four to six bulls per one hundred cows on the open range. The months of gestation for a cow could still be counted on nine fingers. Some ranchers cut the calves away from the cows to wean them. Some, like Hank Cresswell, left it up to the cows to kick their calves off. But management was changing, and for the better. Better quality, purebred bulls were being introduced to upgrade their progeny. Hereford, Shorthorn, and Durham breeds of cattle were popular. Frank Bloom, like Goodnight, was partial to Herefords, which had originated in England, and he was buying them out of Kansas, Iowa, from Conrad Kohrs in Montana, and C. E. Broadbrooks in Canada. In a letter to Survant dated May 30, 1902, Bloom, discussing purchasing bulls, advised: "[I]n buying them I would look for low down blocky chaps. . . ."[6]

When the Bloom Cattle Company moved north, Frank traded the problem of encroaching settlers for one of cattle rustlers and horse thieves. Indians from the neighboring Fort Belknap and Fort Peck reservations stole horses, which they could trade for whiskey. Cattle rustling was a flourishing business.

Although the Great Northern Railroad was completed in 1893 along the Milk River, settlement in this area was slow to follow, partly because of renegade Indians and the likelihood of extreme winter weather. Possibly realizing that settlers would soon be coming over the eastern horizon, the same apprehension that moved Cresswell and Day from South Dakota to Billings and on to Canada, was prompting the Bloom Cattle Company to look north of the border for a more secure land tenure.

In his autobiography, published in 1913, Theodore Roosevelt wrote:

> [T]he homesteaders, the permanent settlers, the men who took up each his own farm on which he lived and brought up his family, these represented from the National standpoint the most desirable of all possible users of, and dwellers on, the soil. Their advent meant the breaking up of the big ranches; and the change was a National gain, although to some of us an individual loss.[7]

In a letter from his office in the First National Bank of Trinidad, Frank wrote to Survant in January of 1902: " We are considering the Canadian proposition and have the business underway. As I wrote you before we have said nothing to anyone except

Cresswell and Day." In February, Frank wrote: "I will enclose a map with Cresswell and Day's lease marked so you can see where they are located. . . . What would you think of one or two townships on White Water and then up on Whitemud nearer Cypress Hills if there is no settlers?" In April he wrote from Trinidad: "I have written Sutherland to gather his 2 yr olds that are not going to calve and we will do the same here. Ought to raise a good herd between ranches. . . . I think thoroughbred bulls go in duty free. Of course we will put thoroughbreds in as we want to make a model herd up there by breeding right and have plenty hay and pastureland to keep up the bulls in winter and have our calves come in the spring."[8]

A letter to Survant shows Bloom's attention to detail in supervising the ranches under his management: "I notice in the last bill 3 cases of coal oil 1st, 5th, & 21st and 6 lbs of whang leather which I suppose is for some of the boys."[9]

The turn of the century was a banner period for cattlemen. The recent worldwide involvement of the United States in the Spanish-American War and the Boxer Rebellion had sparked the economy and created a demand for canned and cured beef for our troops on foreign shores. The weather had been favorable and the prairie of the Great Plains, from Canada to Mexico, was wearing its Sunday best.

The *Medicine Hat News*, Alberta, Canada, February 2, 1905 reported:

> Mr. Thatcher has just returned from a trip to Texas and in that state as well as in Colorado, the cattle have wintered splendidly. Colorado ranches have only had to combat one bad snow storm this winter, and it was of short duration. Bad weather is seldom experienced after the first of February and the cattlemen are consequently in good spirits. Very few losses are reported in the firm's [Cresswell Cattle Co.] Canadian herd this winter, as compared with last and the anticipated increases in prices has imparted a very buoyant feeling in the cattlemen.

Canadian historian Simon M. Evans described the era thus:

> The opening years of the 20th century saw a surge of optimism pass through the cattle industry in western Canada. Vast areas of southern Alberta and Assiniboia were still covered with native grasses. Canadian and British cattle companies continued to flourish using, for the

most part, the extensive methods of the open range. Cattle companies from the United States moved onto the Canadian range with the support of the Government, and to the newcomers "the virgin prairie of southwestern Saskatchewan looked like the promise land."[10]

In 1902, the Bloom Cattle Company expanded from its base in Montana into Canada. The company secured a base acreage of twelve townships from C. E. Broadbrooks along the Whitemud River. The Canadian Mounties kept track of cattle running on Canadian leases, and as the company's cattle from ranges on both sides of the international border were bound to mix, the Canadian Department of Interior required that the company's cattle in Canada be branded differently from those on the Montana range. Bloom had Survant purchase Broadbrooks' Tee Down Bar cattle brand (⏊). Spreading out from Broadbrooks country, the company leased an additional twenty-eight townships. This made a total of forty townships on which to run Tee Down Bar cattle. Their range of 921,600 acres ran from west of the Whitemud and north of the Canada–United States border to the Cypress Hills. Cresswell Cattle Company's lease of forty-two townships for its some twenty thousand Turkey Track cattle neighbored the Tee Down Bar on the east, running between the Whitemud and Swift Current Creek.

Carr Blasingame, cowboy for the Circle Diamond, composed the following tune while on the trail in 1902.

"The Old Circle Diamond"

Way down in Colorado, where mountains are high,
and plains are wide and drear,
Is the Bloom Cattle Company, they buy and sell
And are known both far and near.
They make up many herds an' ship 'em north,
An' they're always mighty fine.
They unload at Billings an' throw 'em on the trail
For the south Canadian line.

Chorus

Then it's "Hold up the leaders, boys, an' drift
Along the drags,
We must graze 'em well each day;
The old Circle Diamonds an' Tee Down Bars,
An' a few old Diamond A's."[11]

The cattlemen's euphoria of 1902 did not last. Not only were dry-land farmers crowding ranchers off the public lands in the northern Great Plains, but the nation's economic upswing during the warring years took a nosedive, and along with it went the cattle market. Grain farming on arable lands on a large scale was possible with new techniques and equipment. Compounding the problems of cattlemen were the shrinkage of live cattle weights and the cost of shipping cattle on the hoof to distant markets, especially to foreign markets.

The ranges in Montana turned dry in 1903 and grew progressively worse during the following three years. The government required dipping and quarantining of transient cattle to combat an outbreak of scabies, thereby adding to the cost of cattle production. The conditions were described by M. E. Miner, a local rancher, in solemn terms:

> The conditions that have prevailed during the past three years, including the present one, have been exactly the reverse of the banner year of 1902. It seems as though nature and man had combined for the undoing of the cattleman. . . . The average price of all the beef steers sold, has not been $3.40 per hundred, and the average net loss to owners cannot have been less that $10 per head when all sources of expense and of loss are taken into account. By that I mean maintaining ranches, winter feeding, interest on investment, taxes, losses by storms, air holes in the river, and to which may be added the tribute levied by the wolves and their counterpart, the irrepressible rustler.
>
> But the day of the range cattle business has passed because of the settling up of the country against which it is folly to contend, and only those who have provided themselves with pastures and winter feed, can remain in the business. . . .[12]

A. E. Cross recommended that weak cattle in his range herd should be put out of their misery "so that available hay could be used to feed the stronger beasts."

The main impact of the winter fell on the shortgrass prairie from Sounding Lake southward to the Cypress Hills and the United States border. Here, losses were estimated at between sixty and sixty-five percent. Even the leading cattle companies, noted not only for their size but also for their skillful and experienced management, lost more than half their herds.

The most obvious effect of the disaster was the failure of a large number of major cattle companies, and the retreat of others to the United States. By 1909 the Turkey Track, the Tee Down Bar, and the Conrad Price Cattle Company had all closed down their Canadian range interests.

After the Circle Diamond and Tee Down Bar closed out their ranches, John Survant continued with his own ranching and business interests. He entered politics and was elected to the Montana State Senate in 1910. Like his fellow fraternity members with the Thatchers' enterprises, Survant was an exemplary man in every aspect. He died in 1951. Survant was elegized by Floyd Hardin in *Campfires and Cowchips:*

> He's restin' now, on the brow of a hill,
> A lookin' down on a valley green,
> Where phantom herds are grazin' still,
> To him a loved, familiar scene.
> Smilin' cowboys, carefree, gay,
> Drift the herd across the plain.
> "All is well," they seem to say,
> "John is ridin' point again!"[13]

EPILOGUE

In 1910, the city directory for Pueblo, Colorado, listed under Live Stock Breeders and Dealers: The Bloom Land and Cattle Company, Cresswell Cattle Company, the Diamond A Cattle Company, and the Hatchet Cattle Company. The address given for the four companies was in the First National Bank Building of Pueblo.

John A. Thatcher passed away on August 14, 1913, fifty years to the day after his arrival in Pueblo. He was seventy-seven. He was survived by his wife, Margaret, and their children, John H., Raymond C., and Lillian. He left an estate valued at $4,111,517.43.[1] Mahlon D., an officer of the other three cattle companies controlled by the Thatchers, succeeded John as president of the Bloom Land and Cattle Company.

In 1968 the Rosemount Estate, with many of its original furnishings, was given to the city of Pueblo by the Thatcher family. John A. Thatcher had the thirty-seven-room mansion built in 1893 at a cost of $60,750. The floor plan is of a twenty-four-thousand-square-foot, three-story home. The estate, which covered a city block, included a six-thousand-square-foot carriage house, and a 2,150-square-foot greenhouse. The buildings are all constructed of pink rhyolite stone. Today, the Rosemount Victorian House is a museum open to the public.

Mahlon D. Thatcher died on February 22, 1916, three years after John, and, like his brother, at the age of seventy-seven. He was survived by his wife Luna and their children, Mahlon D. II, Lydia G., Lucia E., and Luna A. Their three-story Queen Anne-style brick home, Hillcrest, was completed in 1882. After Mrs. Thatcher passed away in 1935, the family gave the mansion to the Red Cross. It was later acquired by the Consistory, a branch of the Masonic Lodge. During his years as a banker, Mahlon was involved financially and as an officer in thirty-two banks. At

Rosemount, home of John A. and Margaret Thatcher. Built in 1891, located in Pueblo, Colorado.

Mahlon's passing, Frank Bloom succeeded him as president of the Bloom Land and Cattle Company.

Frank and Sarah Bloom had four children. Anna, the first-born, died of diphtheria at the age of seven. In 1881, the Bloom family moved into their new three-story brick mansion on East Main in Trinidad, Colorado. Dora Mae, their second daughter, died in 1886 of spinal meningitis. Three years later, their son, Frank H., died of typhoid fever at the age of eleven. The daughter, Alberta G., married William Iliff in 1897. Sarah passed away suddenly in 1928 in Roswell, New Mexico, while accompanying Frank on a ranch inspection trip. Frank held the reins on the Bloom Land and Cattle Company until 1929 when he tendered his resignation because of age and failing health. Frank never recovered from the loss of his beloved Sallie. He had been the man on the ground, the wheel horse for the Bloom Cattle Company, the Diamond A Cattle Company, and the Bloom Land and Cattle Company for fifty-four years. He passed away on Easter Sunday, 1931. He was eighty-eight.

Darwin Daniels, a cowboy and foreman on the Diamond A Ranch near Wagon Mound, New Mexico from 1925 until 1935 said in his book, *The Best 84 Years of My Life:* "Of all the good men

180

Robert O. Anderson purchased the ninety-six-thousand-acre Diamond A ranch in 1960. Courtesy Robert O. Anderson.

I have known in my life, I guess I would put Mr. Bloom on top of the list."[2] Frank's estate was valued at $308,399.69. The Bloom Mansion is a museum open to the public.

Burton "Cap" Mossman succeeded Frank at the helm of the Bloom Land and Cattle Company. John H. Thatcher and Mahlon D. Thatcher II followed their fathers into the ranching business as officers of the company. William Seward Iliff, Frank's son-in-law, came on board as a director.

Good prices for agricultural commodities during and shortly after World War I brought more settlers to file for homesteads on the Great Plains. Acres and acres of prairie, in 160-acre tracts, were turned under with the plow. Falling prices of agricultural products in the late 1920s and the brutal reality that a quarter-section of land on the Great Plains could not provide a living for a family, started an exodus of settlers back to their homeland in the east or on westward to California. The crippling depression and drought of the early thirties was the coup de grâce. John Steinbeck told the story in *The Grapes of Wrath*.

Thousands of acres were laid bare, then abandoned. Persistent winds are always associated with prolonged drought. Old timers contend that "the wind won't stop blowin' till it rains, and it won't rain as long as the wind's blowin'." Much of the heartland of the

Great Plains blew up into mounds of fine soil, burying barbed-wire fences and sod-block hovels.

The era when cattle companies promised investors lucrative dividends and cattlemen grazed their livestock numbering in the hundreds of thousands on the free grass of the Great Plains was fading like a dream at sunrise. The changing times and the bare-bones economics of the livestock industry saw the final chapter written in the books on many of the great spreads.

According to the minutes of a special meeting of the stockholders of the Diamond Cattle Company held in Pueblo on September 20, 1928:

> [T]he meeting was for the purpose of voting on extension of the corporate charter which expired by limitation of law on February 3, 1928 . . . and that since the operations of the Company had shown consistent profits during the past twenty years, and conditions of the cattle industry were now more encouraging than they have been during some of these past years, it appeared to be for the best interests of all concerned to make the renewal.

After Frank stepped down from the presidency, the Bloom Land and Cattle Company and its subsidiary, the Diamond A Cattle Company, continued under the administration of Burton "Cap" Mossman as president and Thatcher heirs as officers and directors. Competing farmers and ranchers were chipping away at the company's large federal and state leases.

In 1928, the Diamond A Cattle Company held the 750,000-acre Diamond A Ranch in South Dakota; five hundred thousand acres in the Circle Diamond and Turkey Track Ranches in Chaves and Eddy Counties; ninety-five thousand acres in the Diamond A Ranch in Mora County, New Mexico; a grazing lease in Montana; and several ranches in Pueblo and Las Animas Counties, Colorado. The Hatchet Cattle Company had spreads in Pueblo, Huerfano and Custer Counties in Colorado, and Grant County in New Mexico.

In 1939, the Diamond A Cattle Company was liquidated and the assets turned over to the Bloom Land and Cattle Company. Real estate owned and leased in New Mexico, South Dakota and Montana and all livestock and physical properties were transferred. In 1944, Leon Williams, a corporate tax consultant, was brought into the Hatchet Cattle Company to advise on tax matters. He held

Entrance to Circle Diamond Ranch, Hondo Valley, New Mexico, circa 1970.

Circle Diamond Ranch, Hondo Valley, New Mexico, circa 1975.

Foreman's house, in Turkey Mountains, Diamond A Ranch, circa 1926. [This was my residence when I managed the Diamond A Ranch in the early 1970s. P.E.P.]

one share of stock. Williams was also employed as a consultant for the Bloom Land and Cattle Company, working on their federal tax problems. In 1944, Williams negotiated to buy all the capital stock of the Bloom Land and Cattle Company. The acquisition included the Diamond A Ranch in South Dakota, the Diamond A, Turkey Track and Circle Diamond ranches in New Mexico. He paid $1,512,000 for the outstanding shares of stock, the deeded land, livestock, ranch equipment, the Diamond A brand, all leased properties and accounts receivable.[3]

In 1945 Williams brought suit against the Commissioner of Internal Revenue claiming a rebate on overpayment of taxes due the Bloom Land and Cattle Company for the years 1940 through 1945. Williams was awarded a sum that returned him a substantial portion of the price he paid for the company.[4]

The February 1945 issue of the *New Mexico Stockman* noted that "John A. McGuire, Lowell, Mass., filed suit in U.S. District Court at Santa Fe against the Diamond A Cattle Company, Roswell, and Leo [*sic*] E. Williams, New York City, as chief stockholder, charging breach of contract for the transfer of $350,000 for sale of the East Side and West Side ranches in Chaves and Eddy counties, and that the alleged failure of the transfer 'was prompted by a larger purchase offer after agreement had been reached.'"[5]

In 1945, the Thatcher heirs bowed out of the Hatchet Cattle Company and Mahlon T. Everhart Sr. took over the corporation. The Hatchet Cattle Company held deeded and leased acreages on the Red Top, Three Rs, Hatchet and Tompkins Mountain Meadow ranches in Pueblo and Custer Counties, Colorado. In New Mexico, the company had a large ranch south of Hatchita and north of the Mexican border.

In 1952, all the leases on the Cheyenne River Reservation that Leon Williams had acquired from the Bloom Land and Cattle Company were canceled. His Diamond A Cattle Company continued operating in South Dakota on deeded property and lands leased from individuals and the government. Leon Williams died in 1958. His ranches were willed to Dartmouth College, his alma mater.

Mahlon T. Everhart, Sr., president of the Hatchet Cattle Company died April 13, 1955 at the age of eighty-two. He had resided in Pueblo for fifty-two years. He was survived by a daughter, Jouett E. Zuendt and two sons, Mahlon T. II, and Jack F.

Burton Mossman retired to his home in Roswell, New Mexico after Leon Williams took over the Bloom Land and Cattle Company. He passed away in 1956 at the age of eighty-nine.

In the early 1950s, Robert O. Anderson, an oilman with headquarters in Roswell, New Mexico, began acquiring ranch properties in New Mexico. He bought the old Bloom Cattle Company's Circle Diamond Ranch headquarters in the Hondo Valley near the present post office at Picacho. In 1960, he purchased the ninety-six-thousand-acre Diamond A Ranch in Mora County, along with the inventory of cattle, horses and the Diamond A brand from Dartmouth College. In 1964, the Anderson family corporation, The Diamond A Cattle Company, bought the Hatchet Ranch south of Pueblo from the Everhart Estate. Anderson continued to add scattered parcels of land along and flanking the Hondo Valley to his Circle Diamond Ranch, almost completely reconstructing the original Bloom Land and Cattle Company's patented holdings. At the crest of Anderson's involvement in the ranching and livestock business he was reputed to be the largest individual landowner in the United States. The Diamond A Cattle Company's cattle, sheep, and horses grazed on over a million acres, scattered from the Big Bend area of Texas to the Hatchet Ranch near Pueblo. His company owned and oversaw feedlots for cattle from Texas to Hawaii. Anderson's Capitol Cattle Company, a commission company, in Austin, Texas, bought and sold thousands of cattle. The slump of oil and cattle prices in the 1960s brought on a gradual dispersal of the Diamond A Cattle Company's real estate and livestock.

The Thatchers and their associates, as pioneering retailers, miners, bankers, and cattlemen, were indeed entrepreneurial leaders on the Great Plains. Thatcher and Everhart descendants, into the fourth generation in the shank of the twentieth century, are still active businessmen, bankers, and ranchers in Colorado and New Mexico.

Unlike many of the large livestock companies that folded during the late 1800s and early 1900s due to the compounding effects of competition for free grass, immigrants moving onto the Great Plains, poor management, adverse weather, and bear markets, the Thatchers and their associates continued on through good years and bad. Their operations were subject to the same weather and market conditions as were all the stockmen on the Great Plains, yet overall, their investments returned great profits. It was not until 1945 that, due to the changing times and the natural forces of attrition, the Bloom Land and Cattle Company ceased to exist.

The fact that the Thatchers, their associates and heirs managed to survive in the ranching business for seventy-four years when others failed was due to the exceptional abilities and loyalties of the personnel and the hands-on management from John and Mahlon in administration, supported by Frank Bloom, Henry Cresswell, Tony Day, Mahlon Everhart, and Burton Mossman.

> May your horse never stumble,
> May your cinch never break,
> May your belly never grumble,
> May your heart never ache.

NOTES

Preface

1. Sandoz, Mari. *The Buffalo Hunters,* 34.
2. Ibid., 95.
3. *U. S. News & World Report,* 19 June 1995, 32.

Chapter 1— Go West Young Man

1. Withers, Geoffrey, *The Thatcher Family of Pueblo, Colorado,* 2.
2. Ibid., 2.
3. Ibid.
4. Fry, Eleanor, "John A. Thatcher, The Genial Banker," 1.
5. Withers, Geoffrey, interview with Mahlon T. Everhart III, 12 July 1972, 1.
6. Haley, J. E., *Charles Goodnight, Cowman and Plainsman,* 206.
7. Fry, Eleanor, "John A. Thatcher, The Genial Banker," 1.

Chapter 2—Don Colora'o

1. February 3, 1984 from Mrs. Alberta C. Iliff Shattuck to Joy Poole.
2. Lavender, David, *The Great West,* 383.
3. Poole, Joy, "The Immortality of Don Colorado," 1986. 4.
4. Bell, William, *New Tracks in North America,* 1867, 90.
5. *Chronicle-News,* 18 August 1929. "Frank G. Bloom, Pioneer Citizen."
6. Iliff Shattuck, Alberta C., "Don Colora'o." 1978.
7. Ibid., 1.
8. Ibid., 2.

Chapter 3—Gold Nuggets and Golden Spikes

1. Lavender, David, *The Great West,* 375.
2. Baker, N. A., "Future of the Plains."

Chapter 4—Trail Drive

1. Lavender, David, *The Great West,* 386.
2. Haley, J. E., *Charles Goodnight, Cowman and Plainsman,* 20.
3. Ibid., 127.
4. Ibid., 129.

5. Ibid., 134.

6. Ibid., 138.

7. *Rocky Mountain Husbandman,* quoting the *Buffalo Livestock Journal,* November 25, 1875.

8. Osgood, E. S. , *The Day of the Cattleman,* 87.

Chapter 5—Free Grass

1. United States Document, *Public Lands,* Chapter XXXIII, December 1, 1883.

Chapter 6—The Dawning of a New Day

1. January 12, 1879 from William E. Graves to J. A. Thatcher, Esq.

2. *A Century in Pueblo, 1871–1971.*

3. *Pueblo Lore,* November, 1992.

Chapter 7—The Associates

1. Fry, Eleanor, "John A. Thatcher, The Genial Banker."

2. Hamner, Laura V., *Short Grass and Longhorns,* 65.

3. *The Medicine Hat News,* February 16, 1905. "The Late H. W. Cresswell."

4. Osgood, E. S., *The Day of The Cattleman,* 46.

5. Thatcher Bros. & Co., Trinidad. Statement dated August 8, 1871 from R. D. Russell, Stonewall, Colorado.

Chapter 8—Black Friday

1. Haley, J. E., *Charles Goodnight, Cowman and Plainsman,* 273.

2. Ibid., 273.

Chapter 9—Trailing to Texas

1. Haley, J. E., *Charles Goodnight, Cowman and Plainsman,* 273.

2. Ibid., 277.

3. Ibid., 280.

4. Ibid., 281.

5. Ibid., 283.

6. Hamner, Laura V., *Short Grass and Longhorns,* 65.

Chapter 10—Beef Bonanza

1. "Pueblo Cattlemen Look Back on Century of Growth."
2. Hayes, A. A., "The Cattle Ranches of Colorado," 891.
3. Gressley, Gene M., *Bankers and Cattlemen*, 49.
4. Dean, Marshall S., *A History of Colorado*, A History of Bent County, 127.

Chapter 11—Looking South

1. Keleher, W. A., *Violence in Lincoln County, 1869–1881*, 3.
2. Ibid., 3.
3. February 6, 1994 from Ellyn Lathan, Curator, Mescalero Apache Tribal Museum, to the author.
4. Bonney, Cecil, *Looking Over My Shoulder*, 135–136.

Chapter 12—West of the Pecos

1. August 25, 1932 from W. T. Thatcher to J. E. Haley.
2. *An Illustrated History of New Mexico.*

Chapter 13—The Bloom Cattle Company

1. Homestead Certificate No. 396, dated April 9, 1881, filed in General land office.
2. Deed, dated July 24, 1883, filed in County of Las Animas, State of Colorado.
3. Osgood, E. S., *The Day of the Cattleman*, 94.
4. Russell, Mrs. Hal, "An Historical Sketch of Las Animas County."
5. Robertson, P. D., and R. L. Robertson, *Cowman's Country: Fifty Frontier Ranches in Texas Panhandle, 1876–1887.*
6. Hamner, Laura V., *Short Grass and Longhorns*, 66.

Chapter 14—Changing Times

1. December 27, 1883 from H. W. Cresswell to John A. Thatcher.
2. W. E. Anderson to Frank Bloom.
3. *The Stockgrower*, November 20, 1886.
4. Pelzer, Louis, *The Cattlemen's Frontier*, 161–2, 180.
5. Betz, Ava, *A Prowers County History.*
6. Goff, R., and R. N. McCaffree, *Century in the Saddle*, 113.

7. Haley, J. E., *Charles Goodnight, Cowman and Plainsman,* 250.

8. Georgia B. Redfield, "The Diamond A Empire," 35.

Chapter 15—White Christmas on the Plains

1. Osgood, E. S., *The Day of the Cattleman,* 102.

2. Robertson, P. D., and R. L. Robertson, *Cowman's Country: Fifty Frontier Ranches in Texas Panhandle, 1876–1887,* 151.

3. Ibid., chap 14.

4. Ibid.

5. April 5, 1930 from E. H. Brainard to J. E. Haley.

6. 1886 from H. W. Cresswell to John A. Thatcher.

7. Osgood, E. S., *The Day of the Cattleman,* 218.

8. Editorial, *Rocky Mountain Husbandman,* August 26, 1886.

9. Hamner, Laura V., *Short Grass and Longhorns,* 68.

10. McCracken, H., *The Charles Russell Book,* 105.

11. Atherton, Lewis, "The Cattle Kings," 210.

Chapter 16—The Turkey Track

1. Hamner, Laura V., *Short Grass and Longhorns,* 65.

2. Ibid., 66.

3. *Canadian Cattleman,* Vol 6/1943, 8.

4. Ibid., 8.

Chapter 18—The Capitalists

1. Atherton, Lewis, "The Cattle Kings," 65.

2. Gressley, G. M., *Bankers and Cattlemen,* 109.

3. Osgood, E. S., *The Day of the Cattleman,* pages 88-89.

4. Atherton, Lewis, "The Cattle Kings," 209.

5. Haley, J. E., *The XIT Ranch of Texas,* 3.

6. Haley, J. E., *Charles Goodnight, Cowman & Plainsman,* 327.

7. Patterson, Paul E., "The Land Grant Issue."

8. Lavender, David, *The Southwest,* 218–219.

9. Betz, Ava, *A Prowers County History,* 282.

10. Goff, R., and R. H. McCaffree, *Century in the Saddle,* 118.

11. Atherton, Lewis, "The Cattle Kings," 210.

12. Gressley, G. M., *Bankers and Cattlemen,* 246.

Chapter 19—Ranch on the Rio Hondo

1. Redfield, Georgia B., "The Diamond A Empire."
2. Shinkle, James D. *Robert Casey and the Ranch on the Rio Hondo*, 163.
3. January 15, 1927 from P. J. White to J. E. Haley.

Chapter 20—Expanding Horizons

1. Fry, Eleanor, "John A. Thatcher, The Genial Banker 1836–1943."
2. Ibid.
3. Raymer, R. G., *Montana. The Land and the People.* Vol. III, 1503.
4. Hardin, Floyd, *Campfire and Cowchips*, 43.
5. *Phillips County (Montana) News*, April, 1931.
6. Hardin, Floyd, *Campfire and Cowchips*, 48.
7. Ibid., 16.
8. Articles of Incorporation, September 30, 1895, Thatcher Archives, Pueblo Library.

Chapter 21—Expanding Horizons

1. Hamner, Laura V., *Short Grass and Longhorns*, 72.
2. Blasingame, Ike, *Dakota Cowboy*, 35.
3. Terrill, Mary, "'Uncle' Tony Day and the Turkey Track," 12.
4. Frank Brown, lease on Lazy F Ranch, 7, copy in author's file.
5. Ward, D. R., *Great Ranches of the United States*, 72.
6. Ibid., 4, 8.
7. Hamner, Laura V., *Short Grass and Longhorns*, 72.
8. Ibid., 73.

Chapter 22—Burton "Cap" Mossman

1. Hunt, Frazier, *Cap Mossman, Last of the Great Cowmen*, 233.
2. Lavender, David, *The Southwest*, 221.
3. Korber, John, "Burton Mossman and the Pueblo Connection," 9.
4. Ibid.
5. Ibid., 10.
6. Jersig, Shelby, "Pioneer Cattleman Stands Tall for Law and Order," 26.
7. Shirley, Glenn, *True West*, October, 1957, 33.

Chapter 23—Ranch on the Cheyenne River Reservation

1. Schell, Herberts S., *History of South Dakota*, 251.
2. Korber, John, "Burton Mossman and the Pueblo Connection," 9.
3. Hunt, Frazier, *Cap Mossman: Last of The Great Cowmen*, 235.
4. Blasingame, Ike, *Dakota Cowboy*, 32.
5. Ibid., 35.
6. Ibid., 35.
7. Jersig, Shelby, "Pioneer Cattleman Stands Tall for Law and Order," 26.
8. Blasingame, Ike, *Dakota Cowboy*, 35.
9. Ibid., 110
10. Ibid., 139.
11. Ibid.
12. Ibid.
13. Minutes of the board meeting of The Bloom Cattle Company, 1907.
14. Ibid.
15. Ibid., note 8, 198.
16. Hunt, Frazier. *Cap Mossman: Last of the Great Cowmen*, 268.
17. Ibid., 273.
18. Ibid., 272.
19. Minutes of the stockholders of The Bloom Land and Cattle Company, December 31, 1938. Original in the Thatcher Archives, Pueblo Library.

Chapter 24—The Hatchet Cattle Company

1. Poole, Joy, Mahlon Everhart III interview, April 30, 1988.
2. Ibid.
3. Articles of Incorporation filed in Pueblo County, October 16, 1902.
4. Keleher, W. A., *The Fabulous Frontier*, 247.
5. Elliott-Fall, Jouett, *Albert B. Fall*, 33.
6. Minutes of directors' meeting, The Hatchet Cattle Company, August 6, 1935.

Chapter 25—Few Equals and No Superiors

1. John Clay quoted in June 1943 issue of *Canadian Cattleman*.
2. Lease dated January 4, 1902. Copy in Bloom Archives, Bloom Mansion, Trinidad.
3. Hamner, Laura V., *Short Grass and Longhorns*, 73.
4. Ibid., 81.

5. Terrell, Mary, "'Uncle' Tony Day and The Turkey Track," 13.

6. Porter, Millie Jones, *Memory Cups of Panhandle Pioneers*, 266.

7. Haley, J. E., *The XIT Ranch of Texas*, 42.

8. Terrell, Mary, "'Uncle' Tony Day and The Turkey Track."

9. Hamner, Laura V., "A Bachelor's Progress" in *Short Grass and Longhorns*.

10. Ibid., 8, 13.

Chapter 26—Circle Diamond and Tee Down Bar

1. Gressley, Gene M., *Bankers and Cattlemen*, 273.

2. Minutes of The Bloom Cattle Company Director's meeting, December 26, 1906.

3. November 30, 1993 from C. G. Barnard, Phillips County Historical Society to the Author.

4. Hardin, Floyd, *Campfires and Cowchips*, 6.

5. Ibid., 6.

6. May 30, 1902 from Frank G. Bloom to John Survant.

7. Roosevelt, Theodore, *Autobiography*, 96.

8. January 13, 1902 from Frank G. Bloom to John Survant.

9. April 6, 1903 from Frank G. Bloom to John Survant.

10. Evans, S. M., *Prairie Forum*, IV Spring 1979, 121.

11. Hardin, Floyd, *Campfires and Cowchips*, 63.

12. November 20, 1905. "The Enterprise," [Montana].

13. Hardin, Floyd, *Campfires and Cowchips*, 78.

Epilogue

1. *Denver News*, December 14, 1913.

2. Daniels, W. D., *The Best 84 Years of My Life*, 139.

3. Court Record, Eagle Butte, South Dakota, Book 72, 753.

4. The Tax Court of The United States, Docket No. 7352.

5. *New Mexico Stockman*, February, 1995, 91.

BIBLIOGRAPHY

An Illustrated History of New Mexico. 1895. Chicago: Lewis Publishing Co.

Atherton, Lewis. "The Cattle Kings." Lincoln, NE: University of Nebraska Press, 1972.

Barker, S. Omar. *Rawhide Rhymes.* Garden City, NY: Doubleday & Co., Inc., 1968.

Bell, William. *New Tracks in North America.* Albuquerque: Horn and Wallace, 1965.

Betz, Ava. *A Prowers County History.* Pueblo, CO: Prowers County Historical Society. 1986.

Blasingame, Ike. *Dakota Cowboy: My Life in the Old Days.* Lincoln: University of Nebraska Press, 1964.

Bonney, Cecil. *Looking Over My Shoulder, Seventy-five Years in the Pecos Valley.* Roswell, NM: Hall-Poorbough Press, Inc., 1971.

Brisbin, James S. *The Beef Bonanza; or How to Get Rich on the Plains, being a description of cattle-growing, and dairying in the West.* With a foreword by Gilbert C. Fite. Norman: University of Oklahoma Press, 1959.

Carlock, Robert H., *The Hashknife, the early days of the Aztec Land and Cattle Company, Limited.* Tucson, AZ: Westernlore Press, 1994.

Daniels, W. D. *The Best 84 Years of My Life.* Wagon Mound, NM: Santa Fe Trail Pub. Co., 1991.

Dean, M. S. *History of Colorado: Five Volumes.* A History of Bent County. Chicago: S. J. Clark Publishing Co., 1918.

Elliott-Fall, Jouett. *Albert B. Fall.* Border Press, El Paso, 1978.

Evans, Simon M. "American Cattlemen on the Canadian Range, 1874–1914," *Prairie Forum* 4 (Winter, 1979), 121–135.

Goff, R. and McCaffree, R. N. *Century in the Saddle.* Denver, CO: Cattlemen's Centennial Commission, 1967.

Gressley, Gene M., *Bankers and Cattlemen.* Lincoln: University of Nebraska Press, 1971.

Haley, J. Evetts. *The XIT Ranch of Texas, And the Early Days of the Llano Estacado.* Norman: University of Oklahoma Press, 1953.

———. *Charles Goodnight, Cowman & Plainsman.* Norman: University of Oklahoma Press, 1949.

Hamner, Laura V. "A Bachelor's Progress." *Short Grass and Longhorns.* Norman: University of Oklahoma Press, 1943.

Hardin, Floyd. *Campfires and Cowchips.* Great Falls, MT: Blue Print and Letter Co., 1972.

Hayes, A. A. "The Cattle Ranchers of Colorado." *Harper's,* 1879. Reprinted Pueblo, CO: Pueblo Regional Library, 1976.

Hunt, Frazier. *Cap Mossman: Last of the Great Cowmen.* New York: Hastings House, 1951.

Keleher, W. A. *The Fabulous Frontier.* Albuquerque: University of New Mexico Press, 1963.

———. *Violence in Lincoln County, 1869–1881.* Albuquerque: University of New Mexico Press, 1982.

Lavender, David. *The Great West.* Boston: Houghton Mifflin, 2000.

———. *The Southwest.* New York: Harper & Rowe, 1980.

McCracken, H. *The Charles Russell Book.* Garden City, NY: Doubleday & Co., Inc., 1957.

Osgood, E. S. *The Day of the Cattleman.* Chicago: University of Chicago Press, 1970.

Pelzer, Louis. *The Cattlemen's Frontier.* Glendale, CA: Arthur H. Clark, 1936.

Pike, Albert. "Narrative of a Journey in the Prairie." *Prose Sketches and Poems.* Boston: Light and Horton, 1834.

Porter, Millie Jones. *Memory Cups of Panhandle Pioneers.* Clarendon, Texas: Clarendon Press. 1945.

Raymer, R. G. *Montana: The Land and the People.* Volume III. Chicago and New York: Lewis Publishing Co., 1930.

Robertson, P. D. and Robertson, R. L. *Cowman's Country: Fifty Frontier Ranches in Texas Panhandle, 1876–1887.* Amarillo: Paramount Publishing Co., 1981.

Roosevelt, Theodore. *The Autobiography.* New York: Scribner & Sons, 1913.

Russell, Charles. *Trails Plowed Under.* Garden City, NY: Doubleday & Co., 1927.

Russell, Mrs. Hal. "An Historical Sketch of Las Animas County" in *A History of Colorado,* Wilbur F. Stone, ed. Chicago: S. J. Clark Publishing Co., 1918.

Russell, Marian. *Land of Enchantment: Memoirs of Marian Russell along the Santa Fe Trail.* Albuquerque: University of New Mexico Press, 1981.

Sandoz, Mari. *The Buffalo Hunters.* Lincoln: University of Nebraska Press, 1978.

Schell, Herbert. *History of South Dakota.* Lincoln: University of Nebraska Press, 1975.

Shinkle, J. D. *Robert Casey and the Ranch on the Rio Hondo.* Roswell, NM: Hall-Poorbaugh Press, 1970.

U. S. Government Publication. *Public Lands.* Chapter XXXIII, Dec. 1, 1883.

von Richthofen, Walter. *Cattle Raising on the Plains of North America.* New York: D. Appleton and Co., 1885.

Ward, D. R. *Great Ranches of the United States.* San Antonio: Ganado Press, 1993.

Suggested Further Reading

Baker, N. A. "Future of the Plains," *Cheyenne (Wyoming) Leader*, May 8, 1868.

Evans, S. M. "American Cattlemen on the Canadian Range, 1874–1914." *Prairie Forum*, Vol. IV, No. 1, 1979.

Evans, S. M. "The End of the Open Range Era in Western Canada." *Prairie Forum*, Vol. IV, 1979.

"Frank G. Bloom, Pioneer Citizen." Trinidad, CO *Chronicle-News*, 18 August 1929.

Fry, Eleanor. "John A. Thatcher, The Genial Banker, 1836–1913." *Pueblo Lore*, November 1992. Pueblo, CO: Pueblo Historical Society.

Hunt, Frazier. *Cap Mossman: Last of the Great Cowmen; with sixteen illus. by Ross Santee.* New York: Hastings House, 1951.

Jersig, Shelby, "Pioneer Cattleman Stands Tall For Law And Order," *New Mexico Magazine*, February, 1993.

Korber, John, "Burton Mossman and the Pueblo Connection," *Pueblo Lore*, April, 1993. Pueblo, CO: Pueblo Historical Society.

"The Late H. W. Cresswell," *Medicine Hat News*, Alberta, Canada, 16 February 1905.

Patterson, Paul E. "The Land Grant Issue," *Livestock Weekly*, Dec. 5, 1985 and Dec. 12, 1985.

"Pueblo Cattlemen Look Back on Century of Growth," *Star Journal Chieftain*, Pueblo, CO. 19 February 1967.

Redfield, Georgia B., "The Diamond A Empire," *New Mexico Magazine*, April, 1942.

Rocky Mountain Husbandman, (Diamond City, White Sulfur, Great Falls, Montana), *1875–1890*. Quoting the Buffalo *Livestock Journal*, November 25, 1875

Rocky Mountain Husbandman, Editorial, August 26, 1886.

Shirley, Glenn. "Cap Mossman and the Apache Devil." *True West*, October 1951.

Terrill, Mary. "'Uncle' Tony Day and the Turkey Track," *Canadian Cattleman*, Vol. 6, June, 1943.

Letters, Minutes, and Manuscripts

Iliff Shattuck, Alberta C. *Don Colora'o.* unpublished manuscript. Bloom Archives.

Letters addressed to or from John A. or Mahlon D. Thatcher are to be found in the Thatcher Archives, Pueblo Library or the Rosemount Museum, Pueblo, Colorado.

Letters addressed to or from Frank G. Bloom are to be found in the Bloom Archives, Bloom Mansion, Pioneer Museum, Trinidad, Colorado.

Letters addressed to James Sutherland are to be found in the Rio Grande Historical Collection. New Mexico State University, Las Cruces, New Mexico.

Letters addressed to John Survant from Frank Bloom are to be found in Montana Historical Society files, Helena.

Letters and research notes to and from J. E. Haley are to be found in the Nita Stewart Haley Memorial Library, Midland, Texas.

Minutes of meetings of The Bloom Cattle Company, Diamond A Cattle Company, and the Hatchett Cattle Company are to be found in the Thatcher Archives, Pueblo Library.

Poole, Joy. *The Immortality of Don Colorado,* unpublished manuscript. Copy in author's file.

Withers, Geoffrey. *The Thatcher Family of Pueblo, Colorado,* unpublished manuscript. From the files of Red Withers, Pueblo, Colorado.

INDEX

C

Greenhorn River, 161
Greensburg, Kansas, 126
Groom, Texas, 165, 166

H

H. W. Cresswell & Co., 37–8, 70,
74
Hachita, New Mexico, xviii
Haley, J. Evetts, 33, 42, 63, 89, 166
Hall, Sallie Monroe, xviii
Hall County, Texas, 139
Hamner, Laura V., 167, 198
Hansford County, Texas, 147
Hansford Land and Cattle Com-
pany, 66, 100, 147–51, 153
Hardin, Floyd, 134, 171–72, 177
Harding County, New Mexico,
167
Harding, President, 162
Hardscrabble Creek, 38
Harper's Magazine, 53
Harris, Franklin and Company,
140
Harris, W. E., 152
Harris-Brownfield Bar W, 161
Hash Knife (Hashknife), 117, 141,
144
*Hashknife, The Early Days of The
Aztec Land and Cattle Company,
Ltd., The*, 144, 198
Hatchet Cattle Company, xv,
160–2, 179, 182, 185
Hatchet Gap, New Mexico, 159
Hatchet Ranch, 159, 161, 163, 185,
186
Head, R. C., 99
Hemphill County, Texas, 88
Henry, Judge J. W., 4
Herd Quitter, The, 136
Hess Wagons, 11
Hillcrest (Thatcher home), 179
Hinsdale, Colorado, 134, 169
Holbrook, Arizona, 117

Hole-In-The-Rock Ranch, 38, 41,
70, 77
Home Creek, 48, 74
Homestead Act of 1862, 25, 28
Hondo River, 66, 81, 123, 128, 140
Hondo Valley, Lincoln County,
New Mexico, 65–6, 83, 8–6,
121, 123–4, 126, 128, 185
Hoover, Austin, xviii
Horsehead Crossing, 21, 57
Hotel Brunswick, Austin, Texas,
77
Howard, Frank, 139
Huerfano County, Colorado, 51,
126, 182
Huerfano Rivers, 77
Hunt, Frazier, 144
Hutchinson County, Texas, 147

I

Ice Age, xiii
Iliff School of Theology in Den-
ver, 113
Iliff, Alberta G. Bloom, xviii, 113,
180
Iliff Shattuck, Alberta, xviii, 7, 12
Iliff, John Wesley, 5, 113
Iliff, William Seward, 113, 180–1
Illinois, 143
Indian Territory in Oklahoma, 74,
100, 137
Indian Wars, 7
Indiana, 35

J

J. E. Haley Memorial Library,
Midland, Texas, xvii–xviii
JA Ranch, 90, 115, 120, 139
JJ Ranch, 56, 114
John Clay and Company, 100, 167
John Deere farm implement com-
pany, 35

Wet Mountain, 159, 163
Wheeler County, Texas, 88
White Mountain, 60, 65, 121
White Oaks, 161
White River Cattle Company, 151
White Water, 174
White, Mahlon T., xvii
White, Phelps, xviii
White Mud (Whitemud) River,
 xv, 101, 174, 175
Wilcox, Arizona, 171
Williams, Leon (Leo) E., 182,
 184–5
Wilson, Billy, 82
Winchester .30.30 rifle, 29, 37, 79
Wister, Owen, 113
Wolf Creek, 74, 90–1, 99
Wood Mountain, 168
Wootton, Richens "Uncle Dick," 9

Wounded Knee Creek, 129, 138
Wyoming, 6, 21, 41, 56, 74, 96, 114
Wyoming Stock Growers' Associ-
 ation, 80
Wyoming Territory, 113

X

XIT ranch, 23, 115, 120

Y

Yellowstone River, 115, 134, 153

Z

Zuendt, Jouett E., 185
Zuni Indians, 60

—